Trading Territories

Trading Territories
Mapping the Early Modern World

Jerry Brotton

CORNELL UNIVERSITY PRESS

Ithaca, New York

For Rachel Holmes

First published in the United States of America
by Cornell University Press, 1998

Originally published in Great Britain in 1997
by Reaktion Books, London

International Standard Book Number 0-8014-3499-8

Designed by Humphrey Stone
Jacket designed by Ron Costley
Photoset by Wilmaset Ltd, Wirral
Colour printed by BAS Printers, Hants
Printed and bound in Great Britain by Biddles Ltd,
Guildford and King's Lynn

Librarians: A CIP catalog record for this book is available
from the Library of Congress.

Contents

Acknowledgements

Thanks go to all the following friends and colleagues who in a variety of ways have helped in the completion of this book: Peter Barber, Maurizio Calbi, Jess Edwards, Bok Goodall, Lorna Hutson, Adam Lowe, Jessica Maynard, Miles Ogborn, Stephen Orgel, James Siddaway, Burhan Tufail, Shobha Thomas, Evelyn Welch and Richard Wilson. Particular thanks go to Roy Porter for encouraging me to publish the project in the first place, and to everyone at Reaktion Books. I am grateful to the School of English at the University of Leeds for providing me with the institutional assistance to complete the current work. I am also particularly indebted to Catherine Delano Smith for offering me intellectual space and learned advice, and Denis Cosgrove for a wealth of astute observations and incisive suggestions on the entire project, as well as generously allowing me access to his own work on globes and globalism. Friends and comrades in South Africa have been a constant source of solace and inspiration, and for this I would like to thank Zackie Achmat, Dhiannaraj Chetty, Mark Gevisser, Jack Lewis and Hugh McLean. Rob Nixon and Anne McClintock helped at crucial times, and have been wonderfully supportive friends and allies. Kenneth Parker appreciated what I was trying to do before virtually anyone else, and I am profoundly grateful to him for his unstinting support and friendship, as well as his wicked sense of humour, which helped me through some of the more difficult moments. Palmira Brummett's brilliant and extremely important work on early modern Ottoman history has been a consistent source of illumination and inspiration to me, and I would like to thank her for all her astute comments and encouragement. Guy Richards Smit watched me complete this book from afar, but still remains 'the wind beneath my wings'. Lisa Jardine has seen this book develop over the last few years, and gave me the courage to breach intellectual boundaries with it at every turn. Her dynamism and intellectual stimulation have been a consistent source of inspiration to me; working with her has been, and continues to be, a truly rewarding experience. Finally, but most importantly, I would like to dedicate this book to Rachel Holmes. Without Rachel, I would never have got round the Cape, in more ways than one.

1 Bernard van Orley, *Earth Under the Protection of Jupiter and Juno*, *c.* 1520–30, gold, silver, silk and wool. Patrimonio Nacional, Madrid.

2 Detail of the terrestrial globe, from illus 1.

3 Martin Behaim, terrestrial globe, 1492, vellum on plaster mounted in metal meridian. Germanisches Nationalmuseum, Nuremberg.

4 Fra Mauro, world map, 1459, ink on parchment. Biblioteca Nazionale Marciana, Venice.

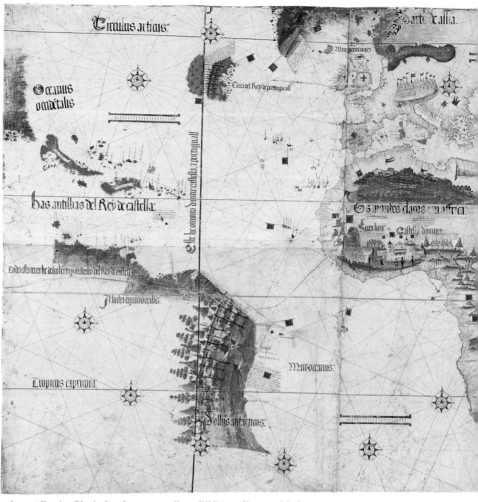

5 Anon, 'Cantino Planisphere', *c.* 1502, vellum. Biblioteca Estense, Modena.

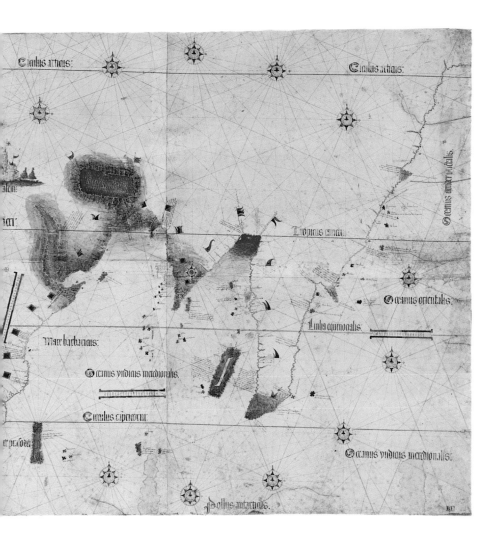

Circulus articus·

Circulus articus·

Oceanus amerioridalis·

alom·

IDI·

Tropicus cancri·

Oceanus orientalis·

Mare barbaricus·

Linha equinoctialis·

Oceanus yndicus meridionalis·

Circulus capricorni·

r pricon·

Oceanus yndicus meridionalis·

Pollus antarticus·

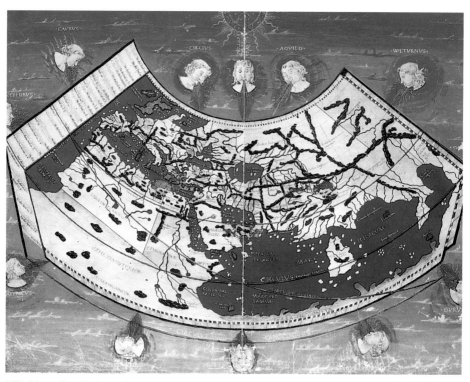

6 World map from Ptolemy, *Geographia c.* 1466, vellum. Biblioteca Nazionale, Naples.

7 Portrait of
Ptolemy from his
Geographia,
c. 1453, vellum.
Biblioteca Nazion-
ale Marciana,
Venice.

8 Anon, *Takiyüd-
din and his
colleagues at work*,
c. 1580, vellum.
Istanbul Univer-
sity Library.

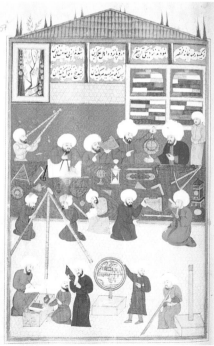

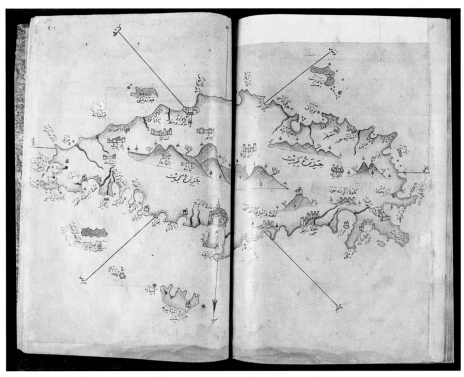

9 Piri Reis, map of Crete, from the *Kitāb-i baḥriye*, 1521, vellum. Topkapi Saray Museum.

1 Introduction

A Wealth of Territory: Mapping the Early Modern World

In 1525 Catherine of Austria married João III of Portugal in a union aimed at cementing an alliance between two of the most powerful royal houses in sixteenth-century Europe. The first was the House of Habsburg presided over by Catherine's brother, Charles V, Holy Roman Emperor, and the second the Portuguese House of Avis, ruled by João III, self-styled 'Lord of Guinea and of the Conquest, Navigation and Commerce of Ethiopia, Arabia, Persia and India'.[1] To commemorate the union João commissioned a series of lavish tapestries entitled *The Spheres* from the renowned tapestry workshops of Brussels. They were a particularly appropriate symbol of the union between João and his prospective wife, who arrived in Lisbon armed with an impressive array of her own collection of Flemish tapestries as well as illuminated manuscripts, assorted family jewellery, furniture and an extensive wardrobe, objects of aesthetic magnificence designed to emphasize her political importance.[2] Attributed to the designs of the Flemish painter Bernard van Orley, the *Spheres* series culminated in a final tapestry depicting a terrestrial globe which, flanked by the Portuguese king and his new wife Catherine, vividly portrayed the territories to the south and east of Portugal to which its sovereign João claimed entitlement (illus. 1).

This remarkable tapestry portrays the two newly married sovereigns as Jupiter and Juno, majestically presiding over the terrestrial and celestial worlds which they survey. The symmetry of the composition, which suspends the terrestrial globe between the sun above and the moon below, enhances the centrality and authority of João and Catherine as rulers of not only the terrestrial sphere but also a celestial sphere from which emanates an image of symmetrical perfection and, by implication, global harmony. However, it is the terrestrial sphere which occupies the symbolic centre of the composition (illus. 2), and it is upon this that João regally rests his royal sceptre, and to this that Catherine gestures with her right hand. Whilst the celestial iconography of the tapestry seeks to confer legitimacy and authority upon the imperial line personified by the union between João and Catherine, it is the terrestrial globe which becomes the primary focus for the ratification of this legitimacy and authority. João's sceptre rests upon the city of Lisbon, the home of the Portuguese House of Avis, since the early fifteenth century the point of departure for Portuguese sea-borne voyages down the west coast of Africa and, from the turn of the sixteenth century, the starting point for maritime initiatives deep into the Indian Ocean. With incredible geographical accuracy the globe triumphantly records in both its nomenclature and its flags flying the arms of Portugal those territories claimed by the Portuguese maritime empire by the time of João's ascent to the Portuguese crown in 1521. Moving south and eastwards from the king's sceptre, the globe portrays West Africa, the Cape of Good Hope, East Africa, the entrance to the Red Sea, the mouth of the Persian Gulf, Calicut in southern India and, at the furthest reaches of contemporary geographical knowledge, on its easternmost extremity, Malacca and the spice-producing islands of the Moluccas, all territories putatively under the sway of João and his new queen Catherine. In one breathtaking visual conceit the globe visualizes João's claim to geographically distant territories, whilst also imbuing his claims with a more intangible access to esoteric cosmological power and authority reflected in the celestial iconography which surrounds the central terrestrial globe.

It is the geographical scope of the *Spheres* tapestry and the ways in which it envisages claims to political power and authority over distant territories which forms the basis of this study of the ways in which princes, geographers, diplomats, sailors and merchants mapped the diverse spaces of the early modern world. The sweep of territory and ocean stretching eastwards through Africa, the Indian Ocean and Southeast Asia, curving back towards Europe through the Red Sea and into the Mediterranean world and Central Asia, is the focus of this analysis, not only in terms of how such geopolitical spaces came to be mapped by the geographers of the early modern world, but also in terms of the ways in which the encounters and exchanges carried out across these territories came to define the space of Europe itself as a social and cultural entity. The most resonant visual and material object which traces this geographical and social process is the map itself, the global geographical representation of the shape of the world as it appeared to sovereigns like João III and Charles V, the changing shape of which is traced in this study.

The history of this mapping of the early modern world inevitably invokes scientific narratives of the development of the discipline of geography as an increasingly objective and definitive account of the relations of points and features on the earth's surface to each other. Historians of both science and cartography have provided illuminating accounts of the part played by fifteenth- and sixteenth-century geography in shaping subsequent scientific traditions which pursued the ideals of objective, verifiable and disinterested enquiry.[3] However, these accounts have invariably imposed a retrospective logic upon the development of cartography and geographical knowledge, which establishes a teleology of progress upon such material where little exists. They have also tended to reify world maps and globes as protoscientific objects divested of any wider social significance, and have limited the ways in which these objects came to be utilized both practically and imaginatively within a range of social situations. It is this utilization of maps and globes which I am concerned with throughout this book.

To take the reproduction of the terrestrial globe within the *Spheres*. The tapestry is itself an apparently unfamiliar context within which to represent such a highly specific image. However, in many ways the reproduction of the globe within it is highly appropriate in terms of the cultural status accorded to maps and globes more generally as valuable material objects which were enhanced by their ability to function at different levels of social life. Like the tapestry itself, they were often utilized by sovereigns such as João III for their ability to ratify their owners' claims to imperial authority, to symbolize, or more often than not speculate on, the size, extent and potential enlargement of specific territorial domains. However, such world maps were not simply tools of imperial administration. Like the *Spheres* tapestry, which mediated a diplomatic *rapprochement* between the royal households of Habsburg and Avis, they also operated within an elaborate diplomatic economy of gift exchange. To proffer the gift of geographically precise and aesthetically magnificent maps of the world established not only the indebtedness of the receiver, but also emphasized the power and authority of the giver, who could subsequently be seen to command an unprecedented level of political authority, purchasing power and access to the arcane skills involved in map production. As with the terrestrial globe at the centre of the *Spheres* tapestry, such maps invariably privileged the territories to which their eventual owner laid claim. They expressed the wonder and excitement of such exotic, commercially tantalizing distant territories whilst also offering a reassuringly sanitized vision of long-distance travel, whose deleterious effects were quietly occluded within lavishly illustrated maps of the known world.

The social value accorded to world maps and globes throughout the late fifteenth and sixteenth centuries was therefore not predicated on their scientific accuracy alone. Whilst they were valorized for their demonstration of learning, they were also valued for their ability to operate within a whole range of intellectual, political and economic situations, and to give shape and meaning to such situations. In his highly influential treatise *De principis astronomiae et cosmographiae* published in 1530 (illus. 10), the Habsburg-sponsored geographer Gemma Frisius was in no doubt as to who benefited from the increasingly fashionable globes of the type reproduced in the *Spheres* tapestry, arguing that:

GEMMA PHRY:
SIVS DE PRINCIPIIS ASTRONO·
miæ & Cofmographiæ, Deçȝ vfu Globi ab eodem
editi. Item de Orbis diuifione, & Infu-
lis,rebufcȝ nuper inuentis.

Veneunt cum Globis Louanii apud Seruatium Zaffenum, &
Antuerpiæ apud Gregorium Bontium fub fcuto Bafilienfi.

10 Frontispiece to Gemma Frisius, *De principis astronomiae et cosmographiae*, 1530. British Library, London.

The utility, the enjoyment and the pleasure of the mounted globe, which is composed with such skill, are hard to believe if one has not tasted the sweetness of the experience. For, certainly, this is the only one of all instruments whose frequent usage delights astronomers, leads geographers, confirms historians, enriches and improves legists [*les legistes*], is admired by grammarians, guides pilots, in short, aside from its beauty, its form is indescribably useful and necessary for everyone.[4]

By 1570 the Elizabethan geographer and man of letters John Dee, in his highly influential *Preface to Euclid's 'Elements of Geometrie'*, was expanding Frisius' claims, locating maps as well as globes within the domestic interiors of Elizabethan England, and noting that:

While, some, to beautifie their Halls, Parlers, Chambers, Galeries, Studies, or Libraries with; other some, for things past, as battles fought, earthquakes, heavenly firings, and such occurrences in histories mentioned: thereby lively as it were to view the place, the region adjoining, the distance from us, and other such circumstances: some other, presently to vewe the large dominion of the Turke: the wide Empire of the Moschovite … Some, either for their owne jorneyes directing into farre landes: or to understand of other mens travailes. To conclude, some, for one purpose: and some, for an other, liketh, loveth, getteth, and useth, Mappes, Chartes, and Geographicall Globes.[5]

What both Frisius and Dee emphasize in these accounts is the importance of the geographical object in facilitating and envisaging the possibility for social and political change across a wide variety of spheres, including juridical, diplomatic, imperial, historical and commercial contexts within which maps and globes increasingly came to play a part. Geographical artefacts were therefore not simply passive reflections of the expanding social and imaginative horizons of so-called 'Renaissance Man'.[6] They were in fact, as Frisius suggests, dynamic depictions of the changing geographical shape of the early modern world, which were also inextricably bound up with its changing diplomatic and commercial contours. It was through both the image of the map and the globe and their existence as valued material objects that the astronomers, historians, lawyers, grammarians, travellers, diplomats and merchants invoked by Frisius made sense of the shifting shape of their world. Geographical images such as the one portrayed on the *Spheres* tapestry mediated a whole range of social issues, from imperial conflict to commercial expansion. Why is it that it is the image of the globe, rather than a madonna or even an emperor, which becomes central to an understanding of the meaning behind the whole composition of the *Spheres* tapestry? Throughout this book I want to emphasize that it is precisely upon the figure of the globe, as both a visual image and a material object, that many of the social and cultural hopes and anxieties of the period came to be focused. For if the development of the terrestrial globe was coterminous with the geographical expansion of the horizons of the early modern world, then the intellectual and material transactions which went into its production were also symptomatic of the expanding intellectual, political and commercial horizons of this world.

Both the *Spheres* tapestry and the globe it depicts were imbued with strikingly new artistic and social values in direct relation to the changing social dynamics of the world which they inhabited. The continually expanding horizons of overseas travel and territorial discovery had ramifications beyond the field of geography. The encounters with those territories depicted in the globe within the *Spheres* produced a whole new range of social and political possibilities, as well as problems. How were such encounters to be incorporated into contemporary systems of knowledge? To what extent did they question the veracity of such knowledge? What were the mechanisms by which sovereigns legitimately laid claim to such territories? How were the immense and complex commercial ramifications of such encounters to be marshalled for maximum financial benefit, and what impact would the experience of these commercial exchanges have on early modern society? These were the sort of questions which the political élites of the period began to ask, and it was in the shape of material objects such as maps and globes that they looked for answers. These artefacts were perceived and circulated as valuable material objects through which the social practices and political contours of the early modern world were given increasing definition, practically, commercially and intellectually. So, as can be seen in the *Spheres*, it is wholly appropriate to read the global image as representative of a ratification of the territorial, diplomatic and by implication imperial claims of João III. In this sense the subject of the depiction of the terrestrial sphere within the tapestry is, iconographically, readily readable. However, the invocation of this sphere and the territories it portrays is at an even deeper level material, the material embodiment of the transactions that were carried out across such territories, transactions which provided the very possibility for the production of this tapestry and the terrestrial globe upon it. The very substance of the materials and knowledges that went into making up the globe stemmed from the territories depicted upon the globe itself and the complex commercial negotiations which went on within them. The silk which made up the bulk of the tapestry arrived in the Brussels tapestry workshops by overland trade routes which stretched through Central Asia and back into China, routes which were being increasingly supplemented by the influx of silk into northern European ports as a result of the establishment of the Portuguese seaborne route to the east, vividly depicted across the tapestry's globe.[7] The silver and gold which gave the tapestry its glittering aura still reached northern Europe primarily via West Africa, Mozambique and Southeast Asia, all territories prominently marked across the globe as being under Portuguese jurisdiction. The brilliant colours which gave the tapestry its compositional definition also emanated from Portuguese transactions throughout Southeast Asia which brought back natural products such as vermilion, indigo, myrobalans, lac, saffron and alum,[8] which revolutionized European dyeing techniques, and hence the threads used in tapestries such as the *Spheres*. The globe situated at the centre of the *Spheres* can therefore be read as an object within which such transactions became both subject and substance, a material and visual cipher which appeared to make all these transactions possible.

What is even more remarkable about the representation of the globe in the *Spheres* tapestry is that it painstakingly reproduced the eastern coordinates of contemporary maps of the Portuguese discoveries in the east, such as the so-called 'Cantino Planisphere', believed to have been produced in 1502 (illus. 4). Like the globe depicted on the *Spheres*, this Planisphere rendered in minute detail the territories encountered by the Portuguese throughout Africa and Asia, and included the coastline of Brazil, discovered by the Portuguese commander Pedro Alvares Cabral in 1500, just prior to the map's completion. Its elaborate finish and highly decorative appearance suggest that it was in fact a copy of the so-called 'Padron Real' held in the *Armazém da Guiné e Indias* (the official repository in the Guinea and India Office) in Lisbon, on which new discoveries were recorded as soon as information was collated in the light of returning expeditions.[9] In November 1502 Alberto Cantino, an agent acting on behalf of Ercole d'Este, the Duke of Ferrara, paid an anonymous Portuguese cartographer twelve gold ducats for the map, which was subsequently smuggled out of Lisbon and installed in Ercole's library in Ferrara.

The story of the Cantino Planisphere is symptomatic of the diverse circulation of maps and, from the first decades of the sixteenth century, globes, throughout the early modern world. Ercole d'Este clearly valued it as perhaps the most comprehensive and up-to-date visualization of the Portuguese seaborne discoveries. However, the possession of an object which more than any other cultural artefact represented geographically and culturally distant phenomena also appeared to gave Ercole access to a deeper, more intangible level of learning and authority inscribed in the Planisphere's mastery of the terrestrial world. As with the depiction of the celestial world contained within the *Spheres* tapestry, comprehensive knowledge of the terrestrial sphere inferred a level of mystical power and authority which legitimated claims to mastery of more esoteric and celestial realms. Possession of decoratively elaborate and aesthetically magnificent maps such as the Cantino Planisphere empowered their owners in making a series of claims to both worldly and other-worldly authority. In the introduction to his version of Ptolemy's *Geographia*, published in Florence in 1482, the Florentine geographer Francesco Berlinghieri (of whom more later) argued that the discipline of geography and its visual manifestation, the world map:

> ... offers divine intellect to human genius, as if it were by nature celestial, demonstrating how with true discipline, we can leap up within ourselves, without the aid of wings, so that we may view earth through an image marked on a parchment. Its truth and greatness declared, we may circle all or part of it, pilgrims through the colour of a flat parchment, around which the heavens and the stars revolve.[10]

Berlinghieri's evocative formulation is remarkably similar to the relationship between celestial and terrestrial spheres which we see in the *Spheres* tapestry, and is symptomatic of the ways in which geographical information, encapsulated in the 'flat parchment' of maps like the Cantino Planisphere, came to be valued for esoteric, as well as more strictly pragmatic, reasons.

Invariably it was the more pragmatic considerations of trade and diplomacy which provided the immediate impetus for the production of maps such as the Cantino Planisphere. The rhetoric of mysterious power and authority which possession of the map enabled its owner to produce (exemplified in Berlinghieri's poetic account of the importance of geography), ultimately found its equivalence in the political and commercial dynamics which fuelled the production of such maps. An artistically cultivated but also politically astute figure like Ercole d'Este was as keen to possess a map such as the Cantino Planisphere for what it could tell him about the extent of Portuguese interpolation into the spice trade of Southeast Asia, as he was eager to possess the more esoteric powers which its mysterious aura conferred upon him. The map itself is studded with references to the markets and commodities of Brazil, West Africa, the Red Sea and India. Above the Portuguese trading stations established in Guinea on the west coast of Africa, the Planisphere notes that from here:

. . . they bring to the most excellent prince Dom Manuel king of Portugal in each year twelve caravels with gold; each caravel brings twenty-five thousand weights of gold, each weight being worth five hundred *reais*, and they further bring many slaves and pepper and other things of much profit.[11]

Near Sumatra a legend reads:

This island called the *Toporbana* is the greatest island found in the whole world and richest in everything, to wit, gold and silver and precious stones and pearls and very big and fine rubies and all kinds of spices and silks and brocades, and the people are idolators and well disposed and trade with foreigners and much merchandise is taken from here abroad and other merchandise which is not in this island is brought in.[12]

Such maps were therefore keys not only to arcane information and esoteric learning, but also to trade routes, market-places and commodities valued by the merchants and financiers of early modern Europe. The aura they exuded was undoubtedly mysterious and other-worldly for many of the people who possessed them; but this aura was also pervaded by a much more acquisitive and commercial ethos with which the development of maps and globes of the period was inextricably entwined.

The first known terrestrial globe was made by the Nuremberg merchant Martin Behaim, who produced his so-called 'erdglobus' in 1492 as a result of his sustained commercial activities along the coast of West Africa whilst based in Lisbon throughout the 1480s (illus. 3). Like the Cantino Planisphere, which appears to be indebted to Behaim for much of its information on geography and trade, Behaim's globe was covered with a profusion of commercial notes on market-places, goods worth purchasing, local trading practices and the movement of commodities which give little doubt as to the audience to which it was addressed. As such maps and globes increasingly became part of the expanding commercial world of the period, they themselves became exchangeable objects caught up in new and highly elaborate forms of commercial transaction. From the end of the fourteenth century maps and sea charts were being exchanged by merchant-cartographers for consignments of pepper,[13] and by the beginning of the sixteenth century possession of a map was often metaphorically and financially compared to the purchasing of the spices, pepper, silk and precious metals to which it appeared to give directional access. In 1592 the English took a Portuguese carrack off the Azores laden with goods from as far afield as China; the English geographer Richard Hakluyt recorded with wonder that the cargo included 'spices, drugges, silkes, calicos, quilts, carpets', not to mention pearls, musk, civet, porcelain, furs and bedsteads. But above all, Hakluyt recorded with delight the discovery amongst the cargo of a map of the Portuguese trading regions:

… inclosed in a case of sweete Cedar wood, and lapped up almost an hundred fold in fine calicut-cloth, as though it had beene some incomparable jewell.[14]

For Hakluyt, the discovery of the precious map and the navigational (and hence commercial) possibilities which it promised inspired even more wonder and excitement than the exotic and precious commodities captured on board the carrack. The map was even more prized for its promise of offering the English themselves access to territories which would yield such fabulous commodities. The value of maps and geographical information such as that described by Hakluyt therefore took on the more strictly commercial ethos of such goods, as the market-place of sixteenth-century Europe grew in size and complexity. Maps and globes visualized the changing world to people like the English who had been relatively peripheral to the great discoveries of the preceding decades, a changing world which, as Hakluyt suggests, was driven by commercial imperatives.

The maps, sea charts and globes of the period performed a dual function. Not only were they displayed as confident imperial symbols of the power and authority of sovereigns like João III and Charles V when laying claim to territories, or even trading in those to which they claimed possession. They were also highly complex objects which were symptomatic of the more pragmatic diplomatic and commercial exchanges between empires. These empires were characterized by their opportunistic existence as trading territories eager to expand the range of their commercial influence within geographically distant places. However, these were far-flung territories over which the empires had little ability to seriously establish anything like the regimes of colonial power and authority which were to characterize the developments of the proceeding European empires of the eighteenth and nineteenth centuries. It becomes clear that the value of the decorative world map and its spherical corollary, the terrestrial globe, developed, as Frisius noted, primarily through the sheer diversity of situations within which they came to give meaning to the social lives of those people who increasingly used them within the worlds of trade, commerce, art, diplomacy and imperial administration.

New Worlds, Old Worlds

Hakluyt's purloined Portuguese map, the similarly appropriated Cantino Planisphere, and the globe represented on the *Spheres* tapestry all exhibit not only the diverse range of situations within which the geographical image as material object comes to achieve its social value. They also have in common a geographical focus on the territories to the east of what we would today call western Europe, the so-called 'Old World' of central Europe, Africa and Asia prior to the 'discovery' of the so-called 'New World' by Christopher Columbus in 1492. This study deliberately focuses on the redefinition of this Old World rather than the impact of the New World. It does so for two specific reasons. Firstly, the impact of the discoveries in the New World on the imagination of early modern Europe was a much slower process than has often been believed. News of these voyages was invariably partial and contradictory, and the territories discovered were for many years perceived as simply extensions of those of Southeast Asia by commentators reluctant to believe in the existence of a distinct fourth continent (Columbus never subscribed to the belief that he had in fact discovered a separate continent). For many early modern sailors, geographers, diplomats, merchants and monarchs the central political preoccupations of the period remained resolutely fixed on the conflict over possession of territories, monopolization of trade routes and establishment of commercial centres throughout the coastal territories of the eastern Mediterranean, West and East Africa, Southeast Asia and the eastern Indian Ocean. Secondly, the narrative of the discovery of the New World resonated within the western imagination in a manner disproportionate to its apparent impact on the culture of early modern Europe. The stories of cultural encounters with the east lack the immediacy and vividness of the originary dramas of discovery offered in the accounts of the New World, partly because these encounters had been ongoing for centuries before the discovery of America. Anglo-American scholars of the early modern period have also, to differing degrees, embraced a discourse of Orientalism which has created and naturalized the assumption of an East, or so-called 'Orient', as a mysterious, exotic, dangerous and ultimately atavistic place. One of the consequences of this critical marginalization has been the elision of the impact of so-called 'eastern' elements on the Renaissance, one of the most crucial and arguably definitive periods of western European civilization. Traditionally the Renaissance, with all its politically loaded connotations of the 'rebirth' of a selectively Aryan set of Graeco-Roman values, transmitted via a reified classical world, has been expunged of any potentially disruptive 'oriental' components or influences.[15]

As a result, the encounters with the east recorded across the surface of the *Spheres* tapestry have subsequently been downplayed in histories of 'The Renaissance', not only because of the complexity of negotiating cultural difference with peoples, ideas and objects which have retrospectively been simply defined as 'oriental', but also because it is not possible to emplot the narratives of encounter, conquest and domination, which have characterized the Renaissance story of the Americas, across the territories depicted on João's tapestry. The encounters recorded by the Portuguese, and subsequently the English and the Dutch, as they sailed down the African coast and onwards into the Indian Ocean, stretch from the fifteenth century to well into the early seventeenth, and indicate a level of negotiation and transaction in a variety of social, cultural, commercial and political exchanges which cannot be adequately explained within the logic of a much later model of the development of the discourse of colonialism. As the locations marked across the globe of the *Spheres* tapestry indicate, the footholds of early European maritime advances throughout the Old World were highly tenuous coastal locations, ports and key towns which invariably defined rapidly chosen points of commercial contact and exchange. Whilst these forms of transaction and exchange preceded later European practices of plantation and colonization, their ethos was significantly different. What needs to be stressed here is not that these early modern transactions did not lay the foundations upon which later practices of colonization developed, but rather that the territories were perceived by the Portuguese and their immediate successors as possessing cultural and commercial agency which has been denied them by subsequent colonial accounts of their historical status. It is in direct opposition to this perception of a sweeping overview of the wider development of European colonial history that this study defines itself.

Furthermore, the intrepid figure of Renaissance Man does not seem to have ventured across such enormous distances primarily in curious pursuit of new horizons of learning and understanding, armed with an innately superior sense of his identity and 'civilizing' mission. On the contrary, he encountered distant territories as potentially lucrative markets for new commodities which Europe sorely lacked, and for which both he and his fellow voyagers could rapidly create an eager demand back home in the marts of northern Europe. Early voyagers soon became aware of the limitations of European commodities; when the Portuguese commander Vasco da Gama landed in Calicut in southern India in 1498, after the first momentous crossing of the Indian Ocean by a western European, his gifts to the Samorin of Calicut were considered so poor next to the spices, silks and precious metals which surrounded the ruler that local Muslim merchants advised him to refrain from offering the gifts, lest the samorin should take offence.[16] Nor was it surprising that the locals were unconvinced by the claims of the sick and bedraggled Portuguese to have come from a great empire across the sea. The Portuguese came from the sea and apparently lived in ships; in India, only those who were so poor that they could not afford their own patch of land would think of going to sea and living on ships.[17] What is so fascinating about the maps and globes which were directly created from such encounters across the space of the Old World is the ways in which they mediated such experiences, in establishing the transactional nature of such cultural encounters (transactions which often provided the very possibility for the production of many maps of the period). Ultimately they also offered a highly sanitized version of the exchanges between trading territories throughout the Old World.

In seeking to address the critical, cultural and geographical occlusion of what has come to be naturalized as 'the East' within the coordinates of a developing global geography, it should be remembered that the geographical antecedents of the geographers of the early modern world lacked any perception of a directional 'east', or even of the very distinction between the geographical and symbolic concepts of 'west' and 'east'. This becomes clear in a brief consideration of the predominant geographical traditions which informed late medieval thought. In attempting to define the shape and scope of geographers' perceptions prior to the voyages and discoveries of the late fifteenth century, it should be recalled that medieval and early modern geographical thinking was incredibly diverse and contradictory, and in no way cohered as a single body of knowledge. It is not within the scope of this study to provide a comprehensive overview of these varied conceptions of geography. However, I would like to examine, albeit briefly, two particularly visible strands of thought which became increasingly definable throughout the fifteenth century, in order to trace the ways in which they conceptualized the geographical coordinates of the classical and medieval worlds, and see how these coordinates were gradually redefined by the maps and globes of the sixteenth century.

The first particularly dominant cartographic tradition which encapsulated the medieval world picture was the so-called 'T-O' map (illus. 11). Historians of cartography have traced its emergence to the Ionian philosophers of the fifth century BC. However, its development as a cartographic image throughout later centuries gradually took on a decidedly Christian

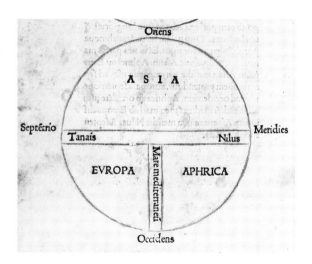

emphasis.[18] Arranged with what might now be called 'east' at the top, the T within the hemisphere as a whole (the so-called 'O') split the earth into Asia (which occupied the entirety of the upper half of the image), Europe (positioned bottom left) and Africa (bottom right). Running from left to right the Tanais (Don) and Nile rivers form the crossbar of the T, whilst the Mediterranean represents the downstroke. A geographical embodiment of the Christian cross itself, this highly schematic map placed Jerusalem at both its geographical and symbolic centre. However, this was a centre which clearly defined itself primarily in terms of the territories which today would be defined as 'eastern', but which were, within the orientation of the T-O map, apparently 'northern'. It was these territories, usually marked on the map as 'Asia', which encapsulated both the hopes and the deep-seated fears of Christian thought and belief. The location of the terrestrial paradise was invariably at the top of such maps and thus at the limit of the known world, which was fixed at the furthest edges of Asia. However, the vast, distant spaces of Asia also yielded their own peculiar terrors. The savage tribes of Gog and Magog, whose emergence on the Last Day would, it was believed, herald the end of the world, were located deep in the region.[19] Within such geographical conceptions, however symbolic they may have been, the space which later came to be defined as 'the East' was clearly a highly significant, imaginatively powerful and symbolically charged place, which gave meaning to the terrestrial as well as celestial world picture of medieval Christianity.

The impact of printing throughout Europe towards the end of the fifteenth century made the simple design of the basic T-O outline a particularly attractive image to reproduce in early books, particularly as it circumvented many of the technical problems associated with unifying images and printed texts. However, the increasing visibility of the T-O image in late fifteenth-century texts failed to reflect the increasing diversification this basic image was undergoing, a diversification which began to emerge in the enormous circular *mappae-mundi* (world maps) of the late fourteenth and fifteenth centuries. These maps were not printed but were lavishly decorated manuscripts, commissioned to adorn the walls of courts and churches. Whilst they retained the directional coordinates of the T-O map, they increasingly registered the proliferation of diplomatic, commercial and territorial information garnered from innumerable travellers who reported on the territories to the directional east of mainland Europe.[20] Even as early as the late fourteenth century such *mappae-mundi* were being produced and circulated as precious objects whose possession was perceived to be coterminous with the precious goods from Africa and Asia to which they seemed to offer access. Whilst still ostensibly adhering to the religious principles underpinning the symbolism of the basic T-O map, evidence which still exists recording the production and reception of these *mappae-mundi* suggests that their central preoccupation was not just with the territories positioned at the top of the map (Africa and Asia), but was also with the exotic, precious goods which came from them. The imposing *mappa-mundi* made for the Portuguese Affonso V in 1459 by the Italian cosmographer Fra Mauro (illus. 4) exemplifies the extent to which the basic outline of the T-O map came to accommodate this proliferation of knowledge about the peoples, customs, goods and market-places in the territories throughout the upper half of the map.[21] Whilst such information was often bizarre and fantastic, it was also at crucial points surprisingly accurate. As well as incorporating recent Portuguese and Italian travels and voyages along the west coast of Africa and into Central Asia, Fra Mauro's *mappa-mundi* also reproduced in some detail the traffic in spices and pepper throughout the Indian Ocean:

Java minor, a very fertile island, in which there are eight kingdoms, is surrounded by eight islands in which grows the 'sotil specie'. And in the said Java grow ginger and other fine spices in great quantities, and all the crop from this and the other [islands] is carried to Java major, where it is divided into three parts, one for Zaiton [Chang-chow] and Cathay, the other by the sea of India for Ormuz, Jidda and Mecca, and the third northwards by the Sea of Cathay.[22]

With a diameter of nearly 2 metres, its parchment surface lavishly covered in details picked out in lapis lazuli and gold leaf, the Fra Mauro *mappa-mundi* combined aesthetic magnificence with highly specific commercial information in a way which was designed not only to confirm the standing of Affonso's court in the Christian world picture, but also to celebrate the growing status of the Portuguese as powerful political and commercial brokers as their voyages of discovery opened new markets down the west coast of Africa. *Mappae-mundi* therefore operated as vast visual encyclopaedias,[23] which offered their patrons a privileged and highly flattering perception of their place in an expanding world, as well as reflecting their particularly acquisitive commercial and political agendas. In the summer of 1399 Baldassare degli Ubriachi, a wealthy Florentine merchant, commissioned four such imposing *mappae-mundi* from two eminent geographers based in Barcelona: Jafuda Cresques and Francesco Becaria.[24] Ubriachi dealt in a range of ivory, precious gems, pearls and jewels purchased through extensive trading connections which he had established throughout the eastern Mediterranean and the Middle East. He intended to take the maps with him on a business trip to sell his precious wares at the courts of Aragon, Navarre, Ireland and England, and to present them to the regents of the provinces in the hope of obtaining rights of free passage and trading privileges throughout their realms. No record exists as to whether or not Ubriachi delivered these expensive and carefully executed maps. However, it seems clear from the records concerned with the commissioning of such *mappae-mundi* that they served a dual function. As well as being magnificent visual encyclopaedias designed to reflect the social prestige of the select few (like Ubriachi) who were able to summon the wealth and intellectual awareness to commission such objects, the maps were perceived by those who commissioned and possessed them to be co-extensive with the precious, rare and unique jewels and gems alongside which, in the case of Ubriachi, they travelled.

Mappae-mundi were to be found throughout the churches and courts of fifteenth-century Europe, vividly mediating the relationship between the celestial and the terrestrial, reflecting not only the piety but also the wealth and worldliness of those who were able to commission such imposing artefacts. However, as the wealth of information which informed the production of *mappae-mundi* such as Fra Mauro's strained the confines of the standard T-O shape, the influence of this image came under increasing pressure with the growing authority in centres of learning throughout Italy of the second predominant geographical tradition which emerged in late fourteenth and early fifteenth-century Europe: Claudius Ptolemy's classical text, the *Geographia*. Ptolemy produced his *Geographia* in Alexandria in the second century AD. Enormous in its scope, it listed descriptions and locations of, and distances between, over 8,000 places known to Ptolemy and his contemporaries, as well as offering an account of how to draw maps of the regions described. Greek copies of the *Geographia* from thirteenth-century Byzantium transcribed its original descriptive chapters and for the first time added maps of the territories described by Ptolemy. By the fourteenth century, Greek copies of Ptolemy's text were arriving in

Italy primarily from Byzantium, along with the works of a whole host of classical Greek and Roman writers such as Plato, Aristotle and Cicero, which were copied and translated by humanist scholars eager to apply the philosophical and scientific lessons of ancient thinking to the rapidly changing world of early modern Europe.[25] By the middle of the fifteenth century luxurious manuscript copies of the *Geographia* were being commissioned by powerful patrons keen to exhibit their taste and erudition in investing in such a revered and increasingly influential classical text.

Although the *Geographia* rejected the symbolic coordinates of the T-O map, it still focused on the supposed 'eastern' territorial sweep of the schematic T-O image. The geographical scope of this world picture, known as the *oikoumene*, stretched from India to the east and Gibraltar to the west, and from the steppes of Central Asia in the north to Ethiopia in the south (illus. 6). However, unlike the medieval *mappae-mundi*, what was significantly different about the emergence of Ptolemy's representation of the known world was its conceptualization in terms of geometrical rather than symbolic principles. The world maps which were subsequently drawn up on the basis of his calculations on the size and shape of the *oikoumene* were therefore emplotted across a predetermined geometrical grid of latitude and longitude whose guiding force was the principles of abstract geometry rather than those of Christian symbolism, which had defined the contours of the T-O map, and, to a lesser extent, medieval *mappae-mundi*. Within this spatially determined grid fifteenth-century geographers were able to plot a proliferation of locations across terrestrial space which was no longer circumscribed by the principles of Christian belief.[26] Ptolemy's impact on the world of geography was to revolutionize a certain perception of space itself, which was no longer charged with religious significance but was instead a continuous, open terrestrial space across which the monarchs and merchants who invested in copies of his *Geographia* could envisage themselves conquering and trading regardless of religious prescription.

The perception of terrestrial space envisaged by Ptolemy's text certainly revolutionized early modern perceptions of geographical space, a legacy which has influenced the discipline of geography to this day. However, the preoccupations which shaped the initial emergence and production of the *Geographia* suggest that it was primarily seized on as a text which offered its predominantly Italian audience an intense and visually arresting engagement with the classical worlds of Greece and Rome. This commerce with the ancients took on both an intellectual and geographical shape, as the scholars and their patrons fashioned the imaginative and topographical landscape of a classical world brought nearer, in both ideal and territorial terms, in the maps produced on the basis of Ptolemy's calculations. Sequential regional maps of his *oikoumene* were carefully reproduced by fifteenth-century illuminators, whose beautifully executed maps offered their owners an insight into the geographical extension of their own contemporary world, mediated through the enduring shape of the Graeco-Roman classical world picture. However, this classical world invoked by the intellectual transactions which saw Ptolemy's text increasingly reproduced in fifteenth-century Italy was, in fact, far more heterogeneous than many commentators have suggested. The frontispiece to the Greek scholar Cardinal Bessarion's magnificently illuminated Greek copy of the *Geographia*, dated around 1453, is indicative of this intellectual heterogeneity (illus. 7). The illustration depicts Ptolemy himself, in the midst of a fabulous landscape of exotic and sumptuous architecture. Richly dressed in an ermine-lined cloak of Islamic design, the ancient geographer holds an astrolabe, an instrument designed to measure the altitude of the sun, moon and stars, and which no self-respecting scholar of geography or navigation would be without. To the right we see Ptolemy's study, represented as a typical fifteenth-century *studioli*, crammed with the paraphernalia of the fifteenth-century geographer. As well as another astrolabe hanging from the wall we can see a quadrant (also used for measuring altitude) and two torqueta (elaborate instruments designed to measure the comparative coordinates of both celestial and terrestrial spheres). Richly bound manuscripts are artfully scattered throughout the *studioli* and, along with the array of scientific instruments, emphasize the esoteric erudition and arcane learning embodied in figures like Ptolemy and his fifteenth-century disciples. The power and authority accorded to such savants is here based on the tangible recovery of the knowledges of the classical world. However, the material substance of this process of intellectual recuperation envisaged in the illustration suggests a more heterogeneous process of cultural transmission. The Islamic influence apparent in Ptolemy's cloak is also reflected in the range of cosmographical and navigational instruments prominently displayed throughout the illustration. Both the astrolabe and torquetum were products of Arabic scientific research which emanated from centres of learning such as Cairo and Seville from the twelfth and thirteenth centuries,[27] as did many of the geometrical and astronomical manuscripts which had such a profound influence on geography and mapmaking in late fifteenth-century Europe – the same sorts of manuscripts that we see scattered throughout Ptolemy's *studioli* in Bessarion's copy of the *Geographia*.

As Bessarion's Ptolemy emphasizes, the *Geographia* did not just define a territorial orientation focused on the east and the geography of the classical world. Its transmission also exemplified the ways in which intellectual and material exchanges between the evolving space of early modern

Europe and a classical world which included Arabic and Islamic influences became vital to the intellectual definition of fifteenth-century Europe. As can be seen from the intellectual and material commerce enacted in Bessarion's copy of Ptolemy, and as with the representations of geographical space in both T-O and *mappae-mundi* traditions, the distinction between a Europeanized 'west' and an Orientalized 'east' is a retrospective divide which makes little sense when trying to trace the cultural exchanges which came to define the shape of the early modern world. The supposedly 'western' world of Europe actually defined itself as coextensive with, rather than in contradistinction to, the classical world of the east, whatever its intellectual and cultural dimensions. Intellectual recognition and respect, political power and authority, commercial wealth and success all appeared to stem from contact with these 'eastern' territories, whose myriad riches appeared to be encapsulated in lavishly illustrated copies of texts such as the *Geographia*. Ptolemy's maps defined the territorial landscape across which such transactions took place. However, this was a landscape which was not only confined to fifteenth-century Italy. As we shall see in more detail, the court of the Ottoman sultan Mehmed II possessed Latin copies of Ptolemy's *Geographia*, as well as actively encouraging further research into the text in Arabic. This visible participation of the Ottoman court in the establishment of the *Geographia* as such a founding text of early modern geography is indicative of the ways in which not only the territories of the so-called east, but also its material wealth and its valuable knowledges were central to the definition of both the geographical and intellectual contours of early modern Europe. What might now be called 'the East' was not a mysterious, distant space for the geographers of the period, but was in fact one from which they drew intellectual and material sustenance. No text more vividly personified this process than the development of Ptolemy's *Geographia*, a text which retained the marks of its hybrid cultural emergence *between* Christian west and Islamic east, rather than defining the social and cultural separation between the two cultures.

It is clearly impossible to neatly separate out the visual and intellectual development of the T-O, *mappae-mundi* and Ptolemaic traditions within the complicated history of late medieval geography. However, it becomes clear that these diverse traditions defined certain geographical parameters within which the mapping of the known world began to achieve a consensual social definition. For all of these geographical traditions, the east was not a separate, mysterious space antithetical to the developing ideals of European civilization. On the contrary, this space which today would be referred to as 'the East' was filled with myriad territories from which early modern scholars imbibed spiritual, intellectual and material sustenance. It is not within the bounds of this study to account for the ways in which this situation changed and redefined itself through the powerfully divisive discourse of Orientalism, which has been so elegantly analysed by Edward Said.[28] However, it could be argued that one particular technological development which hastened the subsequent intellectual and cultural polarization and bifurcation of East and West was the effect of printing on early European society. It is the impact of this new technology on the field of geographical production which should now be considered in looking at the ways in which print began to redefine the geographical shape and social status of maps towards the end of the fifteenth century.

Printing the World

It was a sign of the influence and intellectual authority which manuscript copies of Ptolemy's text had upon the élite of fifteenth-century Europe that it was the *Geographia* which became one of the most prominent texts to be reproduced within the revolutionary new medium of print.[29] In Bologna in 1477 an edition of Ptolemy's *Geographia*, complete with twenty-seven maps, was engraved onto copper plates and printed for the very first time (illus. 12). It was symptomatic of the importance and prestige accorded to the text that the Bologna edition of the *Geographia* was also one of the first printed books to contain engraved illustrations.[30] What this emphasizes is the extent to which both the relatively new discipline of mapmaking and the equally novel medium of print rapidly developed as overlapping technologies throughout the latter half of the fifteenth century. Much has been said about the impact of the invention of printing on literacy and ways of reading, but less has been written about the equally momentous impact of print upon the circulation and communication of visual information within areas such as science, engineering, botany and, of course, geography.[31] Towards the end of the fifteenth century in Europe print allowed for the development of what William Ivins has called 'the exactly repeatable pictorial statement'.[32] Once the cartographic image was transferred to a printing block or plate, it could be repeatedly reproduced, a standardized visual image free from the endless variations and idiosyncratic embellishments which a single illustration could experience at the hands of even the most gifted scribe or miniaturist. The consequences of this innovation, for both the status of maps and their subsequent perception amongst the élites of late fifteenth-century Europe, were profound. Printed editions of Ptolemy like the 1477 Bologna *Geographia* invariably ran to 500 copies which, although still incredibly expensive, began to create a book-buying community across the city-states of western Europe who became increasingly familiar with the Ptolemaic world picture as it

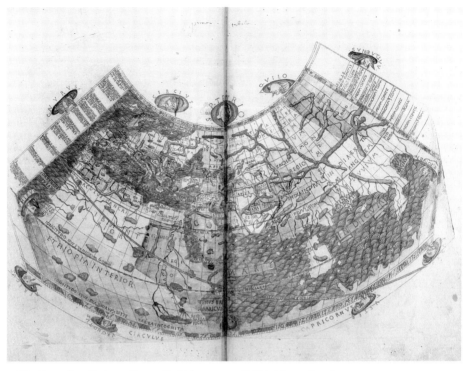

12 World map from Ptolemy, *Geographia*, Bologna, 1477. British Library, London.

extended beyond the confines of its previously highly restricted circulation in manuscript form. The new technologies of print culture also allowed for an increasing diversity and complexity in the field of geographical representation. The advent of both the rolling press and line-engraving on copper plates allowed for not only more uniform and efficient mass reproducibility of images, but also for finer detail in the delineation of the wealth of territories recorded in Ptolemy's work. The use of the technique of copper-engraving in the illustration of maps offered a sharper and more detailed printed impression than the woodcut, and also allowed for detail to be added as new discoveries began to be made. Ptolemy's *Geographia*, with its wealth of place names and precise delineation of coastlines and islands, was not only a prestigious classical text deserving its transmission into the new medium of print. It was also a highly formal technical challenge to the printers and engravers of the new printing presses, who were eager to exhibit their skills in integrating visual information and written text into one carefully crafted book. As a result, they began to establish new vocabularies in their symbolic representation of both land and sea. The subsequent search for standardization in the visual language of cartography allowed for a level of geographical classification, definition and comparison which had been hitherto impossible with the unique and un-reproducible manuscript.

Such classificatory possibilities were increased by the speed with which the medium of print came to dominate the field of geographical production towards the end of the fifteenth century. Within just ten years of its initial publication in 1477, no fewer than six new editions of Ptolemy's *Geographia* had been printed in Rome, Florence and Ulm as well as Bologna, a testimony both to the popularity of the text and the success which the new medium of print had achieved throughout southern and northern Europe. It has been estimated that of the 222 printed maps published up to the year 1500, at least half were directly based on Ptolemy's writings.[33] The impact that this dissemination of his printed text had upon both its community of readers and its status as a precious material object was significant. Whilst wealthy patrons of geography continued to covet lavishly illustrated manuscript volumes of texts such as the *Geographia*, they became increasingly aware of the need to acquire printed editions as well. Possession of a unique manuscript copy of the *Geographia* signified the wealth and power of a patron's ability to commission such an expensive object. However, it rarely moved beyond the confines of his *studioli*, a conspicuous symbol of consumption imprisoned within the confines of its own precious uniqueness. The printed Ptolemy, however, shared a community of élite readers and commentators who built their discourses of geographical awareness around their ability to cross-reference geographical nomenclature which was being increasingly standardized within the latest printed editions. Wealth and conspicuous consumption could be displayed in the possession of a manuscript copy of Ptolemy, but intellectual awareness and politic acumen were even more foregrounded in the study of a printed copy. Wealthy patrons and power brokers such as Federigo Montefeltro and the d'Este family were quick not only to commission new manuscript texts but also to allow their names to be associated with new printing initiatives within the field of academic geography; many, like Mehmed II, ensured that they possessed both manuscript and printed copies of texts such as Ptolemy's *Geographia* in their personal collections.

The establishment of such a widening community of readers also affected the social and cultural status of maps and atlases like the *Geographia*. Whilst the printed copy gradually lost its status as a unique object, its publishers and readers soon realized that such an infinitely reproducible, standardized text produced an imagined community who were beginning to construct their geographical knowledge around their understanding of such printed texts. As the dynamics of print culture began to dictate the intellectual development of geography throughout the sixteenth century, the printers, engravers and geographers of the period began to market their products as valuable commodities which connected a highly disparate collection of scholars, diplomats, merchants and even monarchs scattered across the city-states of early modern Europe. The success of the great cartographers of the mid-sixteenth century, and in particular Abraham Ortelius and Gerardus Mercator, with whom this study concludes, was based as much on their adroit manipulation of the medium of print as it was on their ability to fuse the classical geography of Ptolemy with the latest seaborne discoveries made throughout the period in Africa, Asia and the Americas. In the 'Address' to his highly influential *Theatrum Orbis Terrarum* published in Antwerp in 1570 (illus. 13), Ortelius deftly positioned his comprehensive printed atlas in relation to manuscript and early printed maps:

> For there are many that are much delighted with *Geography* or *Chorography*, and especially with Mappes or Tables contayning the plotts and descriptions of Countreys, such as there are many now adayes extant and every where to be sold; But because that neither they have not that, that should buy them: or if they have so much as they are worth, yet they will not lay it out, they neglect them, neither do they any way satisfie themselves. Others there are who when they have that which will buy them, would very willingly lay out the money, were it not that by reasons of the narrownesse of the roomes and places, broad and large Mappes cannot so be opened or spread, that every thing in them may easily and well be seene and discern'd . . . This I having oft made triall of, I began to bethinke my selfe, what meanes might be found to redresse these discommodities, which I have spoken of, and either to make them somewhat lesse, or, if possibly it might bee, to take them all cleane away. And at length me thought it might be done by that meanes which we have observed and set downe in this our booke, to which I earnestly wish that every student would affoord a place in his Library, amongst the rest of his bookes.[34]

Ortelius had astutely comprehended the increasing limitations of manuscript maps and single-sheet printed maps which, by mid-century, had become not only cumbersome to consult and display, but also gave the purchaser scant return on his investment: such maps were, as a result of the specific conditions under which they had been produced, partial and limited in the range of geographical knowledge which they offered. However, Ortelius' rapidly printed and distributed *Theatrum* could advertise its comprehensive collection of knowledge, drawn from an extensive body of geographers and antiquarian scholars who were united in their relationship to the medium of print and brought together in one highly marketable commodity: the atlas.[35] For a price, as Ortelius tells us, this atlas could take its place amongst the other valuable books required by the sixteenth-century scholar or politician as he negotiated his way through an increasingly diverse world. Ortelius' fellow geographer and

13 Frontispiece to Abraham Ortelius, *Theatrum Orbis Terrarum*, Antwerp, 1570. British Library, London.

engraver Mercator wrote to him in November 1570 in the immediate aftermath of the publication of the *Theatrum* congratulating him on the success of this new printing venture, commenting that:

You deserve no small praise, for you have selected the best descriptions of each region and have digested them in a single manual which without diminishing or impairing the work of any, may be bought for a low price, kept in a small place, and even carried about wherever we wish . . . I am certain that this work of yours will always remain saleable, whatever maps may in the course of time be reprinted by others.[36]

Mercator immediately grasped the fact that with his *Theatrum* Ortelius had produced the ultimate geographical commodity. This was a comprehensive collection of maps and geographical descriptions collected together in one portable book which was reproducible on a mass scale. This was an atlas which was an attractive material object which could be proudly displayed in a library, but which was also made even more valuable and coveted for the specific geographical knowledges it contained. It was these knowledges which, as will become clear, were to be of particular commercial and political use to the diplomats and merchants who eagerly purchased their copies of the *Theatrum*. Like its medieval manuscript predecessor the *mappa-mundi*, the *Theatrum* was a vast geographical encyclopaedia. But if the esoteric aura of mystery and secrecy associated with the *mappae-mundi* appeared to be on the wane with the development of printed atlases such as the *Theatrum*, these atlases were to some extent reinvested with an aura of value equivalent to the access which they seemed to allow to a community of figures of power and influence who increasingly defined their geographical frame of reference through texts like the *Theatrum*.

However, this rapid development of a highly sophisticated printed cartographic tradition did not automatically lead, as has often been assumed, to a complete dismissal of earlier traditions, and in particular those exemplified by the *Geographia*. Both Mercator and Ortelius were instrumental in reshaping the global picture created by Ptolemy into a more recognizably modern image. Gone were the most glaring omissions and erroneous assumptions in his descriptions. The territories of the Americas were painstakingly added, the Mediterranean and the entirety of the landmass of Asia, which had both been greatly overestimated by Ptolemy, were reduced in breadth, and the Indian Ocean was no longer regarded as landlocked. However, geographers such as Ortelius and Mercator remained fulsome in their praise and regard for Ptolemy. In the Preface to his *Theatrum* Ortelius styled him 'the Prince of Geographers', whilst Mercator's homage to the Alexandrian geographer was even more impressive. In 1578 he printed a beautifully engraved edition of the *Geographia* in a series of plates, which remains one of the finest examples of sixteenth-century map-engraving. These responses were undoubtedly carefully executed strategies designed to ensure that the geographical output of the likes of Ortelius and Mercator was perceived to complement rather than reject the thinking of the revered Ptolemy. However, perhaps more significantly, these celebrations of his work reflected an identification of the early printed development of the *Geographia* as a text which opened the possibility for technological innovation and the incorporation of new geographical information unknown in earlier manuscript maps. In this sense, atlases such as Ortelius' *Theatrum* were the logical outcome, rather than intellectual antithesis, of early printed texts of the *Geographia*. These early editions of Ptolemy thrived on technical innovation in print culture as well as gradually assimilating new geographical discoveries. Editions based on the *Geographia*, printed in 1482, incorporated contemporary information into their maps, and by the time of the publication of the Strasbourg edition in 1513 new sheets had been added to the atlas, as well as detailed information on the shape and extent of the African coastline gleaned from Portuguese reports.[37] Rather than an ossified text weighed down by outmoded geographical thinking, these early printed editions became technological and geographical palimpsests, shaped by the experimentations of early printers and geographers, who celebrated the text for its classical authority which allowed it to retain a certain intellectual cohesiveness. This allowed for the possibility of collating an enormous stream of new geographical information within one comprehensible framework, the skeleton structure of the *Geographia*. For this, later geographers such as Ortelius and Mercator were grateful in their attempts to collate the diversity of geographical information into one saleable item: the printed atlas.

However, just as this increasing authority of the printed atlas did not necessarily dispel the authority of Ptolemy, nor did it signal the eclipse of the importance of the unique manuscript map, as has often been believed. Such maps remained significant as the raw material which the more sophisticated and technically polished printed maps produced by the early editors of the *Geographia* and, much later, Ortelius and Mercator drew upon in complementing their classical geographical knowledges taken primarily from Ptolemy. Particularly important in this respect were the so-called 'portolan charts', which were nautical aids used in the art of marine navigation, depicting key ports and basic coastal features of specific stretches of coastline.[38] They were used to assist in the smooth movement of seaborne trade throughout the Mediterranean and the Black Sea, the two areas from which they first emerged towards the end of the thirteenth century. It should be stressed that it was primarily these portolan charts, and not the printed atlases of Ptolemy, which gave merchants and sailors concrete practical information on how to navigate from one place to another. They emerged alongside their topographical counterparts, maps and plans of developing urban centres, such as the anonymous 'Plan of Acre' drawn around 1321 (illus. 14). It was these hydrographical and topographical manuscript maps, charts and plans which became the practical tools used in solving highly particular and local problems such as disputations over land, military surveillance, movement of commercial goods and the maintenance of efficient techniques of irrigation and water supply. As will become apparent in the following chapters, such manuscript maps remained of immense value in the explanation and solution of certain sensitive and politically contentious social situations throughout the sixteenth century. The sheer scale and size of the majority of printed cartography produced during this period ensured that collections of maps such as Ptolemy's *Geographia* and Ortelius' *Theatrum* were invariably of little assistance in practical activities such as maritime navigation and land surveying. However, through their rapid incorporation of a range of geographical traditions (from hydrography to surveying), texts like the *Theatrum* established themselves as the most comprehensive, and hence

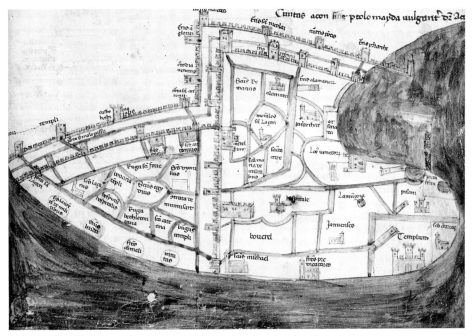

14 Anon, Plan of Acre, *c.* 1321, pen and wash on parchment, from Marino Sanuto, *Liber secretorum fidelium crucis.* Bodleian Library, Oxford.

most desirable, collections of geographical information available. As a result, their construction not only came to define the appearance and presentation of the discipline of geography towards the end of the sixteenth century, but it also shaped the social and geographical perceptions of the political and intellectual élite of the period, perceptions which, as will be seen, were defined by the maps which they consulted.

Conspicuous Cartographies

Although the impact of the technology of print on the field of geography throughout the sixteenth century was profound, print alone cannot be held responsible for the enormous changes which took place in mapmaking and the discipline of geography as they moved into the seventeenth century. A central concern of this book is to stress the interplay between not only print culture and geographical representation, but also the extent to which these fields were themselves inextricably connected to the emergence of new forms of long-distance travel and radically different economic systems of trade and exchange. It is, for instance, impossible to define the extent to which initiatives within cartography and navigation drove the discovery of new long-distance trade routes deep into areas such as Southeast Asia, or vice versa. At certain points it is possible to see the extent to which the discoveries within the fields of printed cartography and navigational knowledge allowed for the swift dissemination of information crucial to the development of long-distance seaborne travel, whilst at others it is possible to see the ways in which the political and commercial exigencies of such travel dictated the production of new forms of geographical knowledge. Whilst recent work on the history of cartography has stressed the centrality of mapping to the territorial claims of western European colonialism, this has been, I would argue, at the expense of an understanding of the ways in which commercial expansion primarily dictated the contours of cartographic production in the early modern period. Whilst claims to territorial possession were undoubtedly imbricated in the rhetoric of early European expansion, the rhetoric often masked the realities of such expansion, which initially involved the highly limited control of extensive sweeps of territory, in favour of establishing strategic points of commercial significance across the map of the early modern world. It is therefore this connection between commercial development and cartographic production which concerns me in the chapters which follow.

The aim of this book is not to produce a comprehensive overview of the incremental mapping of the early modern world, a subject which has been dealt with at great length elsewhere.[39] It is, rather, concerned to investigate certain key moments within this process of global mapping where the interpenetration of print, travel and commerce with mapmaking come together to define the dynamics of geographical representation within this period. An abiding fascination with maps, charts and globes comes from the extent to which such objects, as both texts and visual images, provide their owners with a promise of gain of some sort, be it financial, political or intellectual. The fascination with the geographical artefacts of the early modern period is the extent to which the language of possession which characterizes ownership of the map is so often fused across all three fronts: financial gain, political gain and intellectual gain. In a period where knowledge was invariably, and often highly effectively, conflated with wealth and political power, it is no wonder that the map became such an over-determined object to be possessed and coveted. One final example will suffice to underline this point, and to define the parameters within which this book operates.

In 1594 the Dutch cartographer Petrus Plancius published a map depicting the spice-producing islands known as the Moluccas, located south-east of the Indonesian archipelago. Entitled *Insulae Moluccae* (illus. 15), it is a remarkably accurate hydrographic map, based on Mercator's mathematical projections and incorporating the printing techniques

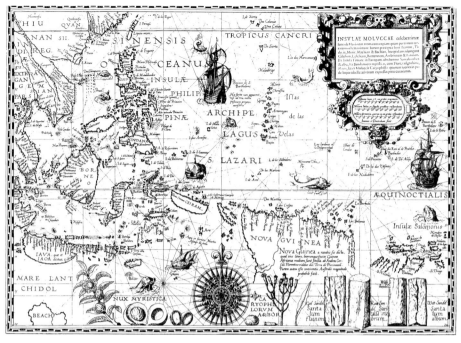

15 Petrus Plancius, *Insulae Moluccae*, engraving, Antwerp, 1594. British Library, London.

developed by Ortelius. Plancius was a respected and enthusiastic supporter of the Dutch joint-stock trading companies who were seeking to break into the lucrative spice market focused on the Moluccas, which had been in the hands of the Portuguese since their arrival in the islands in the second decade of the sixteenth century.[40] Across the bottom of the map Plancius depicted the tangible rewards to be very literally harvested if the Dutch could achieve their goal of full participation in the Moluccas spice trade. Portrayed with botanical exactitude are nutmeg, cloves and three types of sandalwood, all highly lucrative commodities whose influx into Europe had been previously constrained by the increasingly inefficient trading networks operated by the Portuguese. Plancius' engraved map of the islands had been sponsored by commercial interests with substantial financial stakes in ensuring successful Dutch involvement in the trade in spices. Whilst its commissioning and production follow similar patterns to those developed in the production of *mappae-mundi* as well as Ptolemaic and Ortelius atlases, it defines a marked shift in these patterns. This valuable map, a commodified object utilized in the pursuit of financial gain, becomes coterminous with the commodities to which it promises access. Plancius' map is no longer a precious, unique jewel like Ubriachi's *mappae-mundi*, designed to accompany an equally precious consignment of gems, ivory and jewels. Instead it takes its place in the world of goods to which its navigational armature can now allow access. Possession of the commodities on the map becomes, by the end of the sixteenth century, no longer a dream but a tangible reality.

Historically and representationally this book ends with Plancius' map. But what it seeks to trace throughout is the cartographic representations of territory which this map shares with early Christian T-O maps, late medieval *mappae-mundi*, Ptolemy's *Geographia* and João III's *Spheres* tapestry. These are the territories through which the early modern world defined its intellectual and commercial development, the territories adjoining the coastline of Africa as it was traversed by seaborne voyages throughout the late fifteenth and early sixteenth centuries, and the subsequent encounters, by way of the sea, with the Old World of classical thought: the Middle East, Southeast Asia and myriad islands of the Indian Ocean. The following chapters analyse key moments in the history of this mapping of the early modern world, in an attempt to reconsider the impact of encounters with territories to the east of mainland Europe which have been so central, not only to the development of the discipline of mapmaking but also to the idea of European culture itself.

2 An Empire Built on Water: The Cartography of the Early Portuguese Discoveries

The brief analysis offered in Chapter One of the *Spheres* tapestry which adorned the court of João III graphically demonstrates the extent to which the development of the early Portuguese seaborne empire and of cartography as a form of geographical representation were inextricably connected. The profusion of charts, maps and globes which stemmed from the early Portuguese voyages of discovery vividly represented and disseminated information on the scale and extent of Portugal's maritime achievements. A brief comparison of the *Spheres* tapestry with the manuscript maps of Ptolemy's *Geographia* produced throughout Italy only fifty years earlier demonstrates the importance of these Portuguese voyages in redefining the shape of the globe as it was known to the élites of the early modern world. To the modern eye Ptolemy's map appears strange and unfamiliar, with its distended Africa, enlarged Asia and landlocked Indian Ocean. The *Spheres* globe, on the contrary, appears strikingly modern in its accurate delineation of the coastlines of Africa, its depiction of the Cape of Good Hope and its more familiar representation of the islands of the Indian Ocean. The globe stands as an arresting example of the extent to which the Portuguese were responsible for updating global perceptions of the early world, as well as their own changing position within this world. Ptolemy's world picture places Portugal itself on the western edge of the map, whilst the *Spheres* globe vertiginously repositions these coordinates by placing Portugal, with Lisbon at its centre, at the top of the globe, the traditional positioning of the Terrestrial Paradise within the iconography of medieval *mappae-mundi*, a daring representational move indicative of the confidence felt by the Portuguese in the wake of their extensive maritime adventures throughout this period.

Historians of cartography as well as navigation, shipbuilding, armaments and commerce have appreciated the technical innovations which these voyages introduced to their particular fields of specialization as the Portuguese moved across the regions portrayed on the *Spheres* globe. However, more general accounts of the concept of Renaissance Europe (with its heavily moralized connotations of European superiority, integrity, purity and preoccupation with aesthetic beauty) have been reluctant to accord Portuguese exploits throughout Africa and Southeast Asia a place in what has been represented as the pantheon of the glories of the civilization of the Renaissance. Stemming from the nineteenth-century invention of the very idea of 'Renaissance' as a historical epoch, and stretching from critics like Jacob Burckhardt to John Hale, politically and culturally influential definitions of the European Renaissance have tended to ignore not only the extent to which these Portuguese exploits redefined the known space of the early modern world, but also how they transformed a whole range of social activities practised throughout early modern Europe.[1] These ranged from what people wore and what they ate, to how they travelled and how they perceived people of other cultures. This predominantly dismissive perception of Portugal's maritime empire suggests that such critics somehow perceived the Portuguese as devoid of the spirit invariably attributed to Renaissance Man, a spirit characterized by its intellectual curiosity, suspicion of worldly wealth and, above all, its humaneness. Within this purview the Portuguese were seen in their imperial and commercial dealings as essentially pragmatic, ruthlessly pursuing financial gain at the expense of a humaneness which was seen as the cornerstone of Renaissance civility. In many ways this perception of their activities was not wholly inaccurate. By the beginning of the sixteenth century they had established a traffic in slaves along the coast of West Africa which was to define a model of European slavery for centuries to come. This itself has nevertheless been elided within the grand narratives of the Renaissance as a rather unpalatable aspect of the development of the Portuguese empire, rather than as central to the developing economy of early modern Europe.

A far more serious although not unconnected concern for such commentators intent on celebrating a particularly nineteenth-century conception of the Renaissance was the apparent lack of racial purity exhibited in the expansion of the Portuguese seaborne empire. Steeped in contemporary theories of racial classification, highly influential critics of what they termed the 'Age of Discovery', such as J. H. Parry and Boies Penrose, blamed the collapse of the empire towards the end of the sixteenth century on its toleration of miscegenation and racial intermarriage.[2] This led, they claimed, to a process of racial degeneration which eroded the practical and administrative efficiency of the Portuguese abroad. That such beliefs still inform accounts of the rise and fall of the Portuguese empire emphasizes the extent to which the myth of Renaissance Man – purportedly possessed of integrity, curiosity and, above all, racial and cultural purity – still continues to hold sway in the western intellectual imagination. Contemptuously referred to as the 'Kaffirs of Europe',[3] the Portuguese were intuitively perceived to be the antithesis of this version of Renaissance

Man, and as such consigned to the periphery of accounts of 'his' emergence. Predating the Columbian encounters in the Americas by several decades, the early Portuguese seaborne initiatives have been relegated to the status of aberrant voyages, whose motives ran counter to the defining spirit of Renaissance humanism, thus dooming the enterprises to failure.

This chapter rejects such a version of the early Portuguese empire, in the same vein that this book as a whole seeks to contest a version of the Renaissance which has condemned the Portuguese as exorbitant and ultimately degenerate voyagers. Rather than seeing the Portuguese and the motives which underpinned their maritime expansion as antithetical to the ideals of European culture and the intellectual suppositions of Renaissance ideology, it traces the extent to which intellectual and commercial exchanges between Portugal and mainland Europe actually came to define the scope of their expansion. Such an account reorients the place of the early Portuguese empire within the frame of early European culture and society. Perhaps more significantly, it also seriously questions the continuing applicability of the defining values of the Renaissance in explaining the enormous social, political and geographical upheavals generated throughout the world of the late fifteenth and early sixteenth centuries. If the Renaissance ideals of purity, integrity, disinterested enquiry and humaneness begin to look like nineteenth-century inventions designed to recuperate this period in European history, then perhaps today, ironically, it is the Portuguese, with their messy, hybrid histories of commercial, cultural and sexual exchanges with different cultures, who have come to more adequately define the ethos of the early modern world.

One of the most significant spheres within which the Portuguese combined learning and aesthetic beauty with commerce and long-distance travel was within the field of cartography.[4] The massive amount of geographical, hydrographical and navigational data collected from the fleets which were regularly dispatched from Lisbon from the late fifteenth century began to redefine the social status of maps and charts as valuable social and cultural assets. Never before had mapmaking taken on such directly cognitive importance as it did in the face of charting the unknown coastlines and regions encountered by the Portuguese on their voyages down the African coast and onwards into the Indian Ocean. As the wealth of geographical and navigational information grew, so the need for adequate tabulation and representation of such information increased. The result was a gradual shift in the status of maps and mapmaking. The relative swiftness with which the Portuguese moved through the oceans began to redefine the temporality of map production. Older medieval *mappae-mundi* could still reflect the power and prestige of possession of geographical knowledge, but part of their arcane authority emanated from the time spent by learned savants carefully piecing together the encyclopaedic information required to construct their constitutive parts. Prior to the rapidity of the seaborne voyages, the circulation of geographical knowledge, previously slowly transmitted by travellers on foot or horseback, did not require rapid assimilation. However, the Portuguese required more immediate geographical assimilation from their maps, which could portray not only an aesthetic of mystery and splendour in their material presence, but could also rapidly incorporate contemporary discoveries, and hence Portugal's central involvement in changing the basic apprehension of the shape of the globe.

As a result the Portuguese crown employed a collection of professional geographer-navigators towards the end of the fifteenth century who, along with the impact of the new technique of printing, signalled a definable shift in the output of cartography.[5] Often pilots with experience of long-distance seaborne travel, they took on the mantle of learned cosmographical savants. Self-consciously styled as royal cosmographers in the pay of the Portuguese crown, these geographer-navigators retained the aura of learned scholars embroiled in the mysteries of the celestial as well as the terrestrial spheres, whilst wedding their esoteric learning to the practical fields of navigation and mapmaking. As a result, their cartographic output increasingly brought together the practical charts of maritime navigation and the more ceremonial and rhetorical world maps exemplified in the imposing *mappae-mundi* and Ptolemaic maps commissioned by the patrons of fifteenth-century Italy. World maps such as the Cantino Planisphere combined aesthetic magnificence with the hydrographic accuracy provided in the portolan charts which had begun to trace the Portuguese landfalls along the west coast of Africa from the early fifteenth century (illus. 16). Such charts lacked the visual splendour and detailed decoration of world maps like the Planisphere, but they were vital to the painstakingly slow process of mapping previously uncharted coastlines down to the last inlet and harbour. No longer relying on the learned speculation and geographical imaginativeness which characterized so many medieval *mappae-mundi*, maps like the Cantino Planisphere drew on the direct experience of organized long-distance voyages, and the information which they provided. Commercially acquisitive in their primary focus on the location of markets and description of merchandise, such lavish world maps exhibited the influence of geographical and navigational information gleaned from native pilots and geographers whom the Portuguese encountered on their voyages. The material value of such maps was built upon these encounters and exchanges of knowledge, encounters which do not adhere to the classical model of assertive, civilizing and culturally pure Renaissance Man voyaging into the unknown. Nor was their value created

16 Zuane Pizzigano, portolan chart of West Africa, 1424, vellum. University of Minnesota, Minneapolis.

from the realm of a pure and disinterested aesthetic expunged of political or commercial influence. On the contrary, the value of maps like the Cantino Planisphere stemmed from their significance as commercial objects imbued with the ethos of long-distance travel inspired by the search for elusive markets in precious and scarce merchandise, such as the 'seed pearls and grapes and figs and silk and dates and almonds and alum and horses'[6] with which the Planisphere demarcated commercially lucrative cities like Hormuz at the entrance to the Persian Gulf. Combining aesthetic splendour with financial gain, the maps exemplified the commercial and acquisitive impetus of the Portuguese seaborne empire, an impetus which has been interpreted too reductively in classical narratives of the Renaissance. In tracing the production and distribution of these maps, this chapter offers an alternative account of the status of the Portuguese within narratives of the constitution of early modern Europe, and demonstrates the crucial part that mapmaking played in this process.

'Terra Incognita': The Early Portuguese Voyages

Evidence of early fifteenth-century Portuguese involvement with traditional *mappae-mundi* suggests that the crown was well aware of the rhetorical and encyclopaedic value of maps and their claims to universal comprehensiveness. In 1428 the Portuguese prince Dom Pedro returned from an official diplomatic mission which took him to England, Flanders, Germany and Italy. Whilst in Venice the seigniory presented him with a jewel valued at 1,000 ducats, along with a manuscript copy of Marco Polo's travels in the east and a *mappa-mundi*. The sixteenth-century chronicler Duarte Galvão recorded Dom Pedro's travels in his memoirs published in 1563, where he recalled that the prince:

... came home by Italie, taking Rome & Venice in his way: from whence he brought a map of the world, which had all the parts of the world and earth described. The streight of Magellan was called in it The Dragons taile: The Cape of Bona Sperança, the forefront of Afrike, and so foorth of other places.[7]

Galvão clearly superimposed the results of the voyages around the Cape of Good Hope and through the Strait of Magellan upon Dom Pedro's map nearly a century before their actual occurrence in an attempt to emphasize its modernity. Notwithstanding these anachronistic additions, this story of the traffic in decorative *mappae-mundi* between Venice and Portugal underlines a crucial point in the early development of Portuguese apprehensions of cartography. The relationship between jewel, travel, book and map is indicative of the ways in which such precious material objects were perceived by both the Venetian and the Portuguese ambassadorial delegations as constitutive of the imaginative and intellectual processes required to deal with the vicissitudes of unknown, and hence potentially dangerous, long-distance travel. The Venetians were well aware that Dom Pedro had already been active in his sponsorship of Portuguese voyages down the west coast of Africa. The intellectual and logistical apparatuses which they believed were required for such expeditions are vividly encapsulated in the presentation of both book and map, whilst the jewel is tantalizingly evocative of the potentially lucrative rewards which awaited the successful long-distance traveller. Nor were the Portuguese ignorant of the political advantages of surrounding themselves with the vestiges of esoteric geographical learning. Dom Pedro's brother Dom Henrique (better known by his Anglicized name 'Henry the Navigator') had already employed the revered Mallorcan cartographer Jafuda Cresques, one of the mapmakers responsible for the creation of Baldassare degli Ubriachi's *mappae-mundi* in Barcelona in 1400, to provide *mappae-mundi* for the Portuguese court (although little is known of Cresques' subsequent output). In 1457 Henrique's nephew Affonso V commissioned the Italian cartographer Fra Mauro to construct his vast *mappa-mundi*, discussed in Chapter One, which for the first time incorporated details of Portuguese voyages down the west coast of Africa into the relatively homogeneous world picture reproduced in so many late medieval *mappae-mundi*. Completed in 1459, the map was delivered to the Portuguese court upon payment of sixty-two ducats. These appropriations emphasize the extent to which the Portuguese were fully aware of the symbolic power of decorative maps claiming access to universal orders of knowledge. However, the logistical difficulties associated with the increasingly long-distance oceanic voyages which the Portuguese court began to sponsor in the first half of the fifteenth century gradually superseded the highly speculative information which made up such medieval *mappae-mundi*, and led to the development of new forms of cartographic representation.

The bases upon which the Portuguese seaborne empire laid its foundation had from the beginning stemmed from the navigational and hydrographical problems involved in sailing in waters and along coastlines which neither Ptolemy's *Geographia* nor Mediterranean portolan charts were equipped to comprehend. Facing outwards onto a predominantly uncharted northern Atlantic, the Portuguese were in many ways well-situated geographically to initiate a policy of seaborne exploration down the west coast of Africa, and away from the Muslim-dominated territories of North Africa. However, at the same time they were technically ill-equipped to deal with the logistical problems which such exploration involved. Established mapping practices were of little help in these early voyages. Whilst to some extent Ptolemy's geographical data had dealt with Africa, this was confined to hazy regional information which only reached as far as 'Ethiopia' (placed on Ptolemaic maps deep in the southern interior). The western coastline remained uncharted, whilst territory further to the south was simply labelled 'Terra Incognita'. Similarly, the portolan charts used by merchants and sailors throughout the Mediterranean had neither the geographical range nor technical expertise to be of use to Portuguese pilots navigating their way through unknown seas and uncharted coasts. Mediterranean portolan charts conformed to the specifications of conditions unique to that particular sea. Following them invariably involved 'coastwise' sailing, which relied on acquired knowledge of local winds and currents, and the recognition of definable coastal landmarks, markers which were often written into the descriptive accounts which supplemented the charts. Because of the relatively small latitudinal range of the Mediterranean, pilots rarely calculated their latitude whilst sailing, relying instead on 'plain sailing', navigating along a straight line (or 'rhumb') marked on the chart, which linked one coastal location to another.[8] Charts such as Albertin de Virga's 1409 portolan (illus. 17) are typical of those used to assist in this type of navigation. The elaborate network of rhumb lines which crisscross the chart connect the commercial and political world of the Mediterranean, focusing on Venice, Flanders, Alexandria and the Greek islands, carving out a series

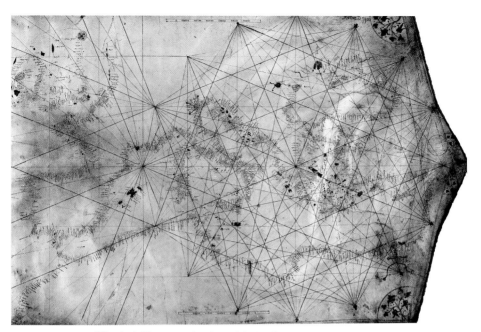

17 Albertin de Virga, portolan chart of the Mediterranean and the Black Sea, 1409, vellum. Bibliothèque Nationale de France, Paris (Charts and Plans).

of definable maritime trade routes along which pilots could navigate with some confidence.

The Portuguese, however, soon realized that this type of navigation, using charts such as de Virga's, was inadequate for the conditions they encountered on their early voyages. The islands of the Atlantic, and the west coast of Africa, offered few readily identifiable landmarks for orientation, whilst sailing in the open sea provided even greater problems. The winds and currents of the Atlantic made coastal sailing a hazardous venture. Yet long-distance oceanic voyages across the natural curvature of the globe made navigation along a straight rhumb line of the type employed in de Virga's chart increasingly difficult. Journeys between two fixed points on the earth's surface had to take into account the curvature of the globe. On short journeys through the Mediterranean this was not too much of a problem, as the relative distortion of the earth's curve produced by a straight line was minimal, and often corrected by location of coastal features. However, this distortion increased across longer distances, such as the Atlantic voyages undertaken by the Portuguese. Pilots increasingly discovered that sailing along a straight line of bearing often left them off-course by several leagues, which made accurate navigation and chartmaking even more difficult. The solution to this problem was the establishment of accurate methods to determine latitude by reference to the sun and the stars whilst at sea and out of sight of land.[9] In so doing the Portuguese could calculate their position in relation to the nearest coastline and hence their relative distance from their point of departure.

By the middle of the fifteenth century commentators began to notice the astronomical developments initiated by Portuguese pilots and cosmographers in the search for reliable methods of oceanic navigation. The German ambassador to the Portuguese court in the 1450s noted in his account of the voyages that there were, 'in addition to the pilots ... master astrologers, very knowledgeable of the routes according to the stars and the Pole'.[10] By the 1470s the Portuguese were extensively using astrolabes and quadrants to determine latitude at sea, and in 1478 the Jewish astronomer Abraham Zacuto circulated his highly influential manuscript the *Almanach perpetuum*. Zacuto's text contained four tables which calculated the position of the sun every day from 1473 to 1476, the coordinates being provided for the hour of noon for the day and place where the calculation was made. Allowing for even greater precision in the establishment of accurate calculations of latitude, such apparently esoteric information was deemed to be of such value that not only was it widely circulated in manuscript form, but it was also increasingly disseminated in print. By the early sixteenth century entrepreneurs had begun to issue solar tables based on Zacuto's calculations. One such entrepreneur was the German merchant and printer Valentim Fernandes. The frontispiece to his *Reportório dos Tempos* (illus. 18), published at his Lisbon printing press in 1518, is indicative of the intellectual and political significance of these tables. The learned geographical savant displays the method of calculating the altitude of the sun to the Portuguese king Manuel I. Dwarfing the two figures is the armillary sphere, signifying the fruitful union between celestial and terrestrial learning and the ultimate symbol of the Portuguese empire itself, a motif which recurs throughout Portuguese iconography of the period, and which was to dominate the design of the *Spheres* tapestry commissioned by Manuel's son João III several years later.

The effect of such relatively sophisticated technical and scientific developments upon maps of the period was significant. The portolan chart drawn by the Portuguese cartographer Pedro Reinel and dated 1483 (illus. 19) shows with startling accuracy the extent of Portugal's maritime activity throughout the Atlantic and along the coast of West Africa by the beginning of the 1480s. The nine Azores islands are all accurately positioned on Reinel's map, as are the Canary Islands, Madeira and the Cape Verde archipelago. But for the first time the map shows with expert precision the location of the four islands of the Gulf of Guinea and, in particular, the island of São Tomé, a pivotal *feitoria*, or trading station, in Portugal's growing commercial network along the west coast of Africa. It

CSeguefeo regimétoda declinaçam do fol pera
per ella faber o mareãte em qual parte efta.f. aquem ou da
lem da linea equinocial.a qual declinaçam he tirada pumtu
almête del Zacuto pello honrrado Gafpar nicolae meftre
fufficiente nefta arte.

CItem faber as que dos.11.dias de março atee do.13. de fe
tembro anda o fol da banda do norte da lineaequinocial.

reflects almost to the month the extent of Portuguese discoveries along this coast, and is dated 1483 on the basis that its information stops at the mouth of the great Congo river which was discovered by the Portuguese commander Diogo Cão early in 1484. Perhaps even more significantly, by 1500 Portuguese chartmakers began to incorporate scales of latitude into their charts of the area. The anonymous chart dated 1500 (illus. 20), which covers much of the territory depicted in Reinel's earlier chart, contains for the first time in the history of cartography a scale of latitude from 18° to 61°, covering the territories from Cape Verde in the south to Flanders in the north.[11] This is the first chart known to incorporate scales of latitude into its overall design, and as such is a striking example of the ways in which, even at this early stage, Portuguese maritime cartography united the symbolics of territorial possession (graphically articulated in the flags which define territorial sovereignty across its surface) with a level of practical navigational precision unparalleled in the history of the early modern mapping of long-distance seaborne travel.

The scientific and intellectual advances made by the Portuguese within the field of astronomy and navigation were central to both the progress of their seaborne voyages along the coast of West Africa and the development of more accurate maps and charts of the territories encountered as a result

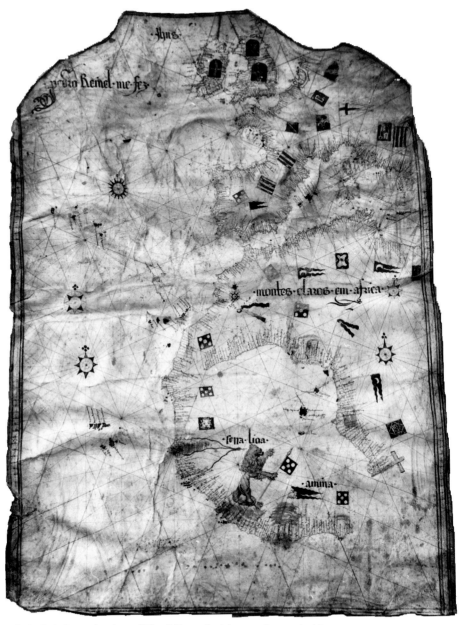

19 Pedro Reinel, portolan chart of West Africa, 1483, ink on parchment. Archives
Départementales de la Gironde, Bordeaux.

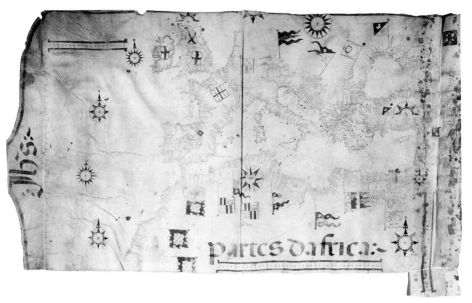

20 Anon, portolan chart of Europe and West Africa, *c.* 1500, vellum. Bayerische Staatsbibliothek, Munich.

of such voyages. However, care must be taken in interpreting these advances within the grand imperial narratives of 'discovery' and 'colonization'. The Portuguese term for discovery, 'descobrimento', was in fact increasingly utilized throughout this period to define the technical achievements of the Portuguese navigational developments as much as the territories to which such achievements provided access.[12] 'Possession' was based not on territories occupied but on the monopolization of logistical and navigational information which established precedence in the creation of commercial sea routes and specific trading points. What primarily motivated these early initiatives was therefore not a desire to 'discover' and colonize territory, but a drive towards commercial expansion. From 1419, when Portuguese ships landed on the Madeira Islands and claimed them for the crown, the fleets dispatched from Lisbon pursued a vigorously commercial agenda. In 1427 ships landed on and subsequently claimed the Azores and by 1434 the Portuguese crown claimed sovereignty over the Canaries. Whilst the term 'conquest' is often applied to Portugal's involvement with these islands, care must be taken when assuming that such conquest simply translated into the terms of territorial possession which characterized later eighteenth- and nineteenth-century European claims to imperial possession throughout Africa and Southeast Asia. Orthodox Christian doctrine stipulated that newly discovered territories populated by peoples unaware of, or antithetical to, Christianity could be justifiably claimed on behalf of a global Christianity. However, such edicts operated as convenient frameworks for early Portuguese exploration rather than as prescriptive imperatives in the name of an all-powerful Christian imperialism. Similarly, the novelty of the supposed conquest of previously unknown territory by sea was very different from the model of territorial 'conquest' of definable territories by land which had preoccupied the armed forces of Christendom throughout the Middle Ages. For the fifteenth-century Portuguese, 'conquest' often signified the establishment of maritime trading routes, the construction of strategic trading stations and the monopolization of specific goods for export, contingent responses to the impossibility of manning larger garrisons and imposing a substantial and systematic colonial administration. Early chronicles of Portuguese involvement with the Atlantic islands repeatedly stress the establishment of the monopolization of the movement of goods to be found in the islands, such as animal skins, fish, wood, sugar and salt. Even when the rationale for such commercial exploits was couched in the language of religious justification, the concept of acquisition and profit remained central to the principles of expansion. In his *Crónica do descobrimento e conquista da Guiné*, completed in 1453, the Portuguese chronicler Gomes Eannes de Azurara noted that Dom Henrique supported the pursuit of involvement in the Atlantic islands for very specific reasons:

For he perceived that no better offering could be made unto the Lord than this; for if God promised to return one hundred goods for one, we may justly believe that for such great benefits, that is to say for so many souls as were saved by the efforts of this Lord, he will have so many hundreds of guerdons in the Kingdom of God.[13]

The language of investment and profit which suffuses Azurara's account is symptomatic of the ethos of Portuguese exploration throughout this period, be it in the traffic in souls, or in the traffic in sugar.

Assisted by the improved astronomical and navigational aids provided by figures like Zacuto, subsequent seaborne expansion was rapid. Averaging over one degree south a year, Portuguese expeditions rounded Cape Bojador in 1434, opening up trade with the Guinea coast. In 1435 the tropic of Cancer was crossed, and by 1460 the Cape Verde Islands had been reached. Between 1472 and 1475 the islands of São Tomé and Príncipe were sighted in the Gulf of Guinea, and in 1473 the equator was crossed by Portuguese ships for the first time. In 1482 the *feitoria* of São Jorge da Mina was established on the Guinea coast, and by 1488 Bartolomeu Dias had rounded the Cape of Good Hope. It was this event which conclusively dispelled the Ptolemaic belief that the Indian Ocean was a landlocked sea, and subsequently opened the way for Portuguese involvement in the proliferation of trading networks which stretched from the eastern coast of Africa to the islands of the Indonesian archipelago. At every point of this remarkable dispersion of the Portuguese seaborne empire, the process of mapping and the expansion of commercial possibilities went hand in hand. In 1469 Affonso V, having already sponsored the production of Fra Mauro's enormous *mappa-mundi* in 1459, awarded a trading monopoly along the Guinea coast to the Lisbon merchant-explorer Fernão Gomes, an agreement described by João de Barros in the account of the Portuguese in West Africa contained in his chronicles *Da Asia*:

As the king was very occupied with the affairs of the kingdom, and was not satisfied to cultivate this trade himself nor let it run as it was, he leased it on request in fourteen hundred and sixty-nine to Fernão Gomes, a respected citizen of Lisbon, for five years, at two hundred thousand reis a year, on condition that in each of these five years he should engage to discover one hundred leagues of coast farther on, so that at the end of the lease five hundred leagues should be discovered, beginning from Serra Leoa ... And among other terms for this contract was that all the ivory should be delivered to the king at the price of one thousand five hundred reis per hundredweight ... Also the traffic of Arguim was excluded, because the king had given it to his son, Prince D. João, as part of his revenue.[14]

Affonso's contract with Gomes united maritime discovery and territorial mapping with the new mechanisms of long-distance trade and exchange required to extract maximum financial benefit from the new commercial possibilities created by Portuguese dealings with West Africa, and in particular the lucrative trade with the ports and islands of Guinea. As well as providing the ivory stipulated in the agreement, the Guinea trade also saw slaves, pepper, musk and gold flowing back into the markets of Portugal. The maps and charts which facilitated this diversification in mechanisms of both trade and merchandise accrued a new social and political status as prized objects in their own right, startlingly vivid material objects which were symptomatic of the impact these new techniques of long-distance travel and commercial acquisition had on the growing prosperity and political importance of both the Portuguese crown and its merchants.

The map was therefore situated at the nexus of these new forms of travel, exchange and acquisition, a point which was stressed in various travel accounts published throughout this period on the nature of the Portuguese incursions along the coast of West Africa. In 1455 the Venetian merchant Alvise Cadamosto took a share in a crown-sponsored initiative to trade along the coast of Senegal. Upon landing there and trading with the locals, he noted of them that:

They were also struck with admiration by the construction of the ship, and by her equipment – mast, sails, rigging, and anchor ... They said we must be great wizards, almost the equal of the devil, for men that journey by land have difficulty in knowing the way from place to place, while we journeyed by sea, and, as they were given to understand, remained out of sight of land for many days, yet knew which direction to take, a thing only possible through the power of the devil. *This appeared so to them because they do not understand the art of navigation, the compass, or the chart.*[15]

Cadamosto's comments say more about his own wonder at the benefits of the compass and the chart than that supposedly exhibited by the local inhabitants. For a merchant like Cadamosto, used to the plain sailing methods adopted in the waters of the Mediterranean, the ability to sail in open sea for extended periods of time using charts and celestial observations must have seemed as fabulous as the goods which he eventually shipped back to Venice via Lisbon. Clearly aware of the advantages of these new developments in the fields of navigation and cartography, Cadamosto did not only return to Venice with the commodities for which he traded along the coast of Senegal. He also arrived back in his native city in 1463 with maps and hydrographic information which he supplied to the Venetian cartographer Grazioso Benincasa. In 1467 Benincasa produced an atlas of portolan charts, including a remarkably up-to-date chart of the coastline of West Africa which reproduced Cadamosto's written observations on his journey down the coast, point by point (illus. 21). Whilst lacking the opulent decoration of earlier *mappae-mundi*, charts like Benincasa's began to move along the same trade routes as the goods to which they provided access. Lacking the esoteric quality of the earlier *mappae-mundi*, they nevertheless established an aura of value precisely in terms of the potential commercial benefit they could provide for their owners. As a result, such charts took their place within a rarefied marketplace peopled by merchants, commercially minded sovereigns and chartmakers, who were no longer in search of the geographical speculation provided by *mappae-mundi* but were instead searching for specific hydrographic and commercial information with which to launch long-distance trading ventures. As a result of the early Portuguese voyages down the west coast of Africa, the perception of marine cartography changed for ever. Mapmaking became tied to the exigencies of commercial expansion, which came to define the shape of early modern cartography as it began to chart systematically the territories to the south and east of mainland Europe over the following decades.

The Myth of a Renaissance Man: Henry and His Navigators

Cadamosto's voyage down the coast of Senegal had been sanctioned by Dom Henrique, on condition that Cadamosto paid half of any subsequent

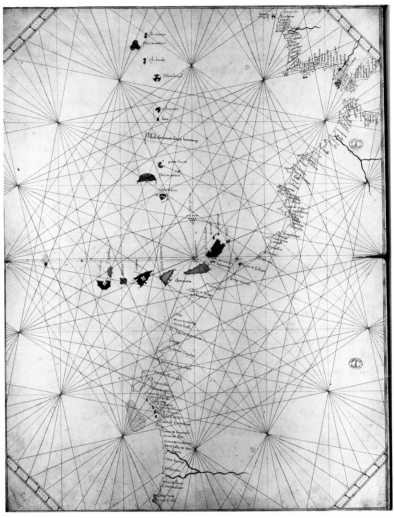

21 Grazioso Benincasa, portolan chart of West Africa and islands, 1467, vellum. Bibliothèque Nationale de France, Paris (Charts and Plans).

profits from his venture to Henrique. Cadamosto's encounter with the Portuguese prince has been seen as just one of the many examples of the apparently enlightened pioneering spirit of Dom Henrique. Historians of the so-called 'Age of Discovery' have enshrined him, above all other Portuguese figures, as an archetypal Renaissance Man, grandly dubbing him 'Henry the Navigator' (illus. 22).

Supposedly exhibiting the classical virtues of piety, learning and patriotic pride, Dom Henrique has been seen as the one potentially recuperable figure in the history of early Portuguese maritime expansion, singlehand-

22 Dom Henrique (Prince Henry the Navigator), detail from Nuno Gonçalves, panel of the Infanté, from the *Veneration of St Vincent*, c. 1471–81, oil on panel. Museu Nacional de Arte Antiga, Lisbon.

edly responsible for not only the spread of the spirit of humanist enquiry and scientific learning in Portugal, but also the organized drive to find a navigable sea route to India. He has been portrayed as being, between 1415 and his death in 1460, the driving force behind the Portuguese voyages down the coast of Africa. The myth of Dom Henrique as the erudite scholar-prince was primarily established by nineteenth-century Portuguese historiography, which was anxious to transplant the Burckhardtian model of Renaissance Man into the history of the early Portuguese seaborne empire, in an attempt to emphasize Portugal's *missào de civilizacào* in the face of anti-colonial struggles against Portuguese authority.[16] Claims that Henrique established an academy devoted to the study of astronomy and navigation at Sagres on the southern coast of Portugal, where he surrounded himself with some of the greatest scientific minds of the period, are unsubstantiated by any contemporary chronicles that recount his life and career. The outlandishness of such claims as to Henrique's learned wisdom stretched as far as some commentators confidently asserting that he read the work of the German astronomer Regiomontanus, whose work was printed over twenty years after Henrique's death! The prince's reputation primarily rests on the admiring account offered in Azurara's *Crónica do descobrimento e conquista da Guiné* (although even here no mention is made of his scientific interests). However, Azurara appears to have deliberately painted him in a positive light as a result of his loyalty to King Affonso V in 1449, when his brother Dom Pedro briefly challenged Affonso for the crown. As Azurara's chronicle was written under the patronage of the king, it is hardly surprising that Dom Henrique is portrayed as a wise and loyal prince.

There is no doubt that Dom Henrique was at the forefront of sponsoring Portuguese voyages into the Atlantic and down the coast of Africa. However, his motives for supporting such expeditions were much more pragmatic than has previously been believed. Rather than seeking to combat the Muslim 'infidel' and convert 'heathens' to the virtues of Christianity, the chronicles of the period indicate that Henrique was in fact far more interested in establishing his own commercial control of the territories claimed by the early Portuguese voyages. As early as 1439 he secured a grant from King Duarte giving him commercial control over the Madeiras, and by the 1440s, as Azurara notes, his thoughts had turned to the Canary Islands:

In the year 1446 the Infant [Dom Henrique] began to make ready his ships to return to the said conquest [of the Canaries], but before doing aught in the same, he requested the Infant Dom Pedro, his brother, who at that time was ruling the kingdom in the name of the King, to give him a letter forbidding all the subjects of these realms from daring to go to the Canary Islands, to war or treat of merchandise, without the command of the said Infant. This letter was granted him, and beside this he was privileged to enjoy a fifth whatever should be brought from there; and this was very rightly given him, considering the great expense which that noble Prince had incurred in the matter of the said conquest.[17]

Earlier in the same year Henrique had secured rights to all trade beyond Cape Bojador on the west coast of Africa, and by the time of his death in 1460 he had established commercial monopolies on fishing, soap and dye-making throughout the Atlantic islands and on the African coast. A good example of his commercial preoccupations emerges from a grant he issued to a prominent member of his retinue, Bartolomeu Perestrelo, who informed him of the commercial possibilities offered by the islands of Madeira and Porto Santo. In 1446 Henrique drew up a grant allocating Perestrelo certain rights to Madeira, of which the prince was to take a cut. They give an idea of what he saw as the benefits of settling such an apparently insignificant island:

Furthermore, it is my pleasure that the said Bartolomeu Perestrelo have for himself all grain mills in the said island, which I hereby give him in charge; and no one but he or whoever he permits may construct mills, except hand mills which may be built by anyone who wishes, but not for the purpose of milling for others, nor may anyone build a horse or water-powered mill except he himself or whoever has his approval. Item: it is my pleasure that he may collect water rights of one mark of silver or its equivalent yearly from everybody or two planks every week of those customarily sawed, paying to us, however, the tenth of all such lumber, in the same way as is paid for other products of the said sawmill, and that Bartolomeu Perestrelo have also the same rights from any other mill that may be built there, excluding veins of iron ore or other metals. Item: it is my pleasure that he have all baking ovens that are equipped with vats ... Item: it is my pleasure that if he has salt for sale nobody but he may sell salt ... Furthermore, it is my pleasure that he have the tenth of everything that is owed to me as income from the island.[18]

However effective such a decree proved to be in practice, it established a model of commercial monopolization which was to characterize Portuguese mercantile activity for the next hundred years.[19]

The picture of Henrique's exploits which emerges from contemporary sources indicates that, rather than being a confident, crusading Renaissance Man, he deftly turned the logistical difficulties of long-distance seaborne travel to his own highly specific, and deeply single-minded, commercial ends. Henrique realized that political power and authority did not lie in claiming huge tracts of ungovernable territory, but in the strategic and highly pragmatic commercial monopolization of key ports and islands. The limited number of maps and charts which emerged from these initiatives were therefore not the result of the flowering of a dynamic humanist impulse in Portugal at the behest of the learned Henrique. They were, in fact, logistical objects facilitating the commercial development of the prince's own network of trading interests.

Nor were these trading interests limited to the singular expansion of the Portuguese economy. They had ramifications well beyond Portugal. As Dom Henrique's commercial agreement with the Venetian merchant Cadamosto makes clear, the prince was fully aware of the need for financial diversification in the development of trade throughout the Atlantic islands and along the coast of Africa in order to maximize the territories' economic potential. The Portuguese crown had neither the capital nor the commercial experience required to capitalize on these new, if geographically distant, markets. Dom Henrique therefore fostered commercial exchanges with mainland Europe, and in particular the trading ports of Italy, in an attempt to draw financial investment into the African and Atlantic trades. In 1447 the Genoese merchant Antoine Malfante drew up an agreement with Henrique which allowed him to trade around the mouth of the Niger river, and in 1455 his fellow Genoese merchant Antoniotto Usodimare was granted rights to trade along the coast of Senegal.[20] These commercial exchanges between mainland Europe and Portugal can be traced in the surviving charts of the period, which indicate the extent to which the commercial developments of the Portuguese discoveries were rapidly assimilated into a wider European network of finance and exchange. In 1448 the Venetian cartographer Andrea Bianco drew a portolan chart which depicted for the very first time Portuguese navigation beyond Cape Bojador (illus. 23). However, it was not completed in Lisbon. The author's legend states that Bianco finalized the chart in London, on his way to Flanders from Venice, having already stopped en route in Lisbon. Bianco's journey became a standard trade route followed by merchants from southern Europe, who were keen to invest in Portuguese voyages down the coast of Africa and out into the Atlantic, and shipped goods traded in these locations to the marts of northern Europe and, in particular, London and Flanders. Bianco's chart imparts the news of the Portuguese discoveries and connects them to the trade routes linking Italy, Lisbon, Africa and Flanders in one continuous line of coastal territory. Whilst his chart was not necessarily produced for commercial navigation, it does, along with several other similar ones of the period, seem to have been the type used by pilots navigating along these trade routes with the commercial support of the Portuguese crown and, more specifically, Dom Henrique. The commercial agreements Henrique made with Venetian and Genoese merchants established a structure of commercial exchanges which defined the expansion of the Atlantic and African trade well into the sixteenth century. By the 1470s Florentine merchants such as Benedetto Dei were trading deep into the interior of West Africa with Portuguese consent; and by 1479 Flemish merchants like Eustache de la Fosse were establishing trading links with the Portuguese *feitorias* in the Gulf of Guinea.[21] As a result, by the end of the fifteenth century a third of all sugar production coming out of Madeira was being exported to the Low Countries, with the revenue from such trading rights going straight into the coffers of the Portuguese crown.[22]

Dom Henrique was not the ideal Renaissance Man of the Burckhardtian imagination, nor was he steeped in the learned mysteries of science and astronomy. Critical attempts to foreground his cultural achievements as an enthusiastic patron of scholarly activity, over and above his drive towards financial expansion and commercial diversification, have sought to separate the pursuit of scholarship from the pursuit of commercial gain. Henrique's career (and the range of intellectual and technological initiatives which emerged within Portugal around this period) emphasizes that, particularly for the early Portuguese voyagers, no such simplistic distinction existed. The process by which the advances made by the Portuguese throughout the fifteenth century were mapped across the charts and global maps of the period is symptomatic of how scholarship

23 Andrea Bianco, portolan chart of Europe and West Africa, 1448. Biblioteca Ambrosiana, Milan.

and commerce were mutually supportive activities in defining the early modern world picture as it emerged as a result of the voyages into the Atlantic and down the coast of Africa. If Dom Henrique can be said to be responsible in any way for the momentous changes which occurred in this world picture, it was through his attempts to fill the gaps in a changing commercial market rather than those on the unfolding map of the early modern world.

Maps and Monopolies: The Portuguese in the Indian Ocean

By the 1480s the growing success of the commercial ventures initiated by Henrique allowed for increased expenditure on voyages seeking a navigable sea route from Portugal to the spice and pepper markets of Southeast Asia, whose merchandise slowly and expensively trickled into mainland Europe via the Red Sea and overland trade routes through Ottoman-controlled Asia. By 1485 João II confidently announced in his oration to Pope Innocent VIII delivered in Rome:

... the by no means uncertain hope of exploring the Barbarian Gulf, where kingdoms and nations of Asiatics, barely known among us and then only by the most meager of information, practice very devoutly the most holy faith of the Saviour. The farthest limit of Lusitanian maritime exploration is at present only a few days distant from them, if the most competent geographers are but telling the truth. As a matter of fact, by far the greatest part of the circuit of Africa being by then already completed, our men last year reached Prassum Promontorium, where the Barbarian Gulf begins, having explored all rivers, shores and ports over a distance that is reckoned at more than forty-five hundred miles from Lisbon, according to very accurate observation of the sea, lands, and stars.[23]

Like Dom Henrique before him, João's account of Portuguese navigational achievement was consistently couched in the rhetoric of Christian expansionism, as befitted the audience to whom it was addressed. Yet the emphasis in João's oration is quite clearly placed on announcing the geographical and technological extent of the Portuguese discoveries, as they exceeded the limits of Ptolemaic knowledge, reaching people 'barely known among us', and stretching as far as the 'Prassum Promontorium', the most southerly point reached by the Portuguese in 1484. However, the one cartographic artefact which came to encapsulate the extent of the Portuguese discoveries prior to the successful passage around the Cape of Good Hope in 1488 was a terrestrial globe designed not by a Portuguese geographer, but by a merchant from Nuremberg.

In 1484, the same year that João composed his oration on the extent of the Portuguese discoveries in Africa, Martin Behaim, a cloth merchant from Nuremberg, arrived in Lisbon, ostensibly to capitalize on the commercial incentives which had been established by Dom Henrique for foreign merchants keen to exploit the trading possibilities offered along the African coast. He subsequently claimed to have sailed with the Portuguese commander Diogo Cão as far south as the tropic of Capricorn between 1482 and 1484. Little is known of Behaim's putative exploits in Portugal and West Africa, but his experience was clearly impressive enough for the burghers of Nuremberg to commission him to produce a terrestrial globe upon his return to the city in 1490 (illus. 24). Behaim was approached by George Holzschuher, a prominent financier who held monopolies on silver mining throughout the region. As a businessman involved in the trade in precious metals, it is not surprising that Holzschuher should have taken such a keen interest in Behaim's first-hand account of the burgeoning gold trade which had sprung up along the west coast of Africa. Behaim's claims to commercial involvement and navigational expertise throughout this region suggested that he possessed the requisite skills to produce a material object worthy of the investment which the burghers of Nuremberg were prepared to make in the terrestrial globe, as well as being able to provide the city's merchants with the valuable commercial

24 Detail of Europe and the Near East from illus 3.

information they required to break into the lucrative trade along the African coast.

Detailed documentation recording the production of the globe indicates that it was a costly and time-consuming commission. Behaim initially drew up a basic manuscript *mappa-mundi* design as a working model from which to complete his globe. This map was designed to a highly decorative standard, as it subsequently adorned the walls of the Nuremberg council chamber. Assisted by an array of painters and craftsmen, Behaim spent over two years supervising the production of the globe, which was subsequently completed sometime prior to his return to Lisbon in July 1493. Estimates calculate that it cost the Nuremberg council the equivalent of £13 17s which, considering the amount of work involved, was a particularly good price for what was to prove to be the first known terrestrial globe made in early modern Europe.[24]

The final dimensions of the globe were no less impressive than the trouble taken in its commissioning and construction. With a diameter of 50 cm, mounted on a stand and richly illustrated in a range of colours, it retains to this day the imposing appearance it must have possessed at the time of its unveiling. The globe contained a proliferation of visual as well as written information (in Latin). It amassed a total of 1,100 place names and depicted 48 flags (10 of them Portuguese), 15 coats of arms and 48 miniatures of kings and local rulers. Daring in its size and scale, Behaim's globe also dramatically displayed the full extent of Portuguese voyages down the coast of Africa, showing evidence that he was at least aware of Bartolomeu Dias' voyage round the Cape of Good Hope in 1488. On Behaim's globe Africa is no longer depicted as a continuous landmass running into Asia, as portrayed in early printed editions of Ptolemy's *Geographia*. Instead, southern Africa is boldly portrayed as a navigable point offering access to an open Indian Ocean. Not only was this the first strikingly secular terrestrial globe depicting the momentous changes which the traditional Ptolemaic world picture was experiencing, but it was also one of the first definitively global geographical images to place Africa as central, rather than peripheral, to the political and commercial world of early modern Europe.

What is most striking about the production of the globe is the extent to which Behaim's apparent commercial expertise in distant places allowed him to acquire the status of geographical specialist, rather than vice versa. For both the Portuguese crown and the Nuremberg burghers, it was this that enhanced his prestige as a geographer and, ultimately, globemaker. Shortly after his arrival in Lisbon in 1484, Behaim had been invited to join the *Junta dos mathematicos*, a group of Portuguese pilots and cosmographers assigned the task of developing more accurate calculations of latitude through solar observation. His intellectual qualifications for such a post appear to have been little more than a dubious claim to have studied briefly under the German astronomer Regiomontanus. What seems to have motivated the Portuguese in offering Behaim such a position was his commercial knowledge, rather than his intellectual credentials. Similarly, the burghers of Nuremberg seem to have been more interested in commissioning a terrestrial globe from the highly partial standpoint of a German merchant versed in long-distance travel, rather than from an experienced pilot or cosmographer more familiar with the arcane mysteries of cosmography and geography. The good burghers were presumably not disappointed with the detailed information which covers the globe. In an extensive legend, adjacent to the meticulous delineation of the west coast of Africa, Behaim managed not only to provide a primarily commercial report of the area, but was also astute enough to insert his own identity into the narrative:

In the year 1484 after the birth of Christ our Lord, the Serene King John II of Portugal caused to be fitted out two vessels called caraveli, which were found and armed for three years. These vessels were ordered to sail past the columns which Hercules had set up in Africa, always to the south and towards sunrise, and as far as they possibly could. The above king likewise supplied these vessels with various goods and merchandise for sale and barter. The ships also carried 18 horses with costly harness, to be presented to Moorish Kings where we thought fit. We were also given various examples of spices to be shown to the Moors in order that they might understand what we sought in their country. Thus fitted out we sailed from the port of Lisbon in Portugal to the Island Madeira, where the sugar of Portugal grows, and through the Fortunate Islands and those of the savage Canarians. And we found Moorish Kings to whom we gave presents, for which they gave return presents. And we came to the kingdoms of Gambia and Golof, where grow the grains of paradise, and which are distant 800 German miles from Portugal. And subsequently we arrived in King Furfur's country [Benin], which is 1,200 leagues or miles distant, and where grows the pepper called Portugal pepper. And far from there we came to a country where we found cinnamon growing.[25]

Offering hydrographic information, details of merchandise to be purchased and the etiquette of pragmatic exchanges of goods with Muslim rulers, Behaim's globe was not a document of colonial territorial expansion but a highly complex representation of a network of territories, ports, goods and people involved in an evolving commercial world.

Nor were these lines of commercial exchange which crisscrossed the globe limited to the trade along the coast of Africa. East of the southern tip of Africa the nomenclature appended to the territories marked for specific attention indicated the extent of Portuguese commercial ambitions once a navigable sea route to Southeast Asia, or the 'Indies', had been found. Deep in the Indian Ocean Behaim placed a lengthy legend detailing with remarkable accuracy the commercial distribution of spices throughout the region:

Item, be it known that spices pass through several hands in the islands of oriental India before they reach our country.
1. First, the inhabitants of the island called Java Major buy them in the other islands where they are collected by their neighbours, and sell them in their own island.
2. Secondly, those from the island Seilan, where St. Thomas is buried, buy the spices in Java and bring them to their own island.
3. Thirdly, in the Island Ceylon or Seilan they are once more unloaded, charged with Customs duty, and sold to the merchants of the island Aurea Chersonesus, where they are again unladen.
4. Fourthly, the merchants of the island Taprobana buy the spices there, and pay the Customs duties, and take them to their island.
5. Fifthly, the Mohammedan heathen of Aden go there, buy the spices, pay the Customs and take them to their country.
6. Sixthly, those of Cairo buy them, and carry them over the sea, and further overland.
7. Seventhly, those of Venice and others buy them.
8. Eighthly, they are again sold in Venice to the Germans, and customs are paid.
9. Ninthly, at Frankfurt, Bruges and other places.
10. Tenthly, in England and France.

11. Eleventh, thus at last they reach the hands of the retail traders.

12. Twelfthly, those who use the spices buy them of the retail dealers, and let the high customs duties profits be borne in mind which are levied twelve times upon the spices, the former amounting on each occasion to one pound out of every ten. From this it is to be understood that very great quantities must grow in the East and it need not be wondered that they are worth with us as much as gold.[26]

With an astute eye to the complexities of long-distance trading practices, Behaim had quickly grasped the lucrative nature of the Asian spice trade. Spices were a relatively light and easily transportable commodity, with a potentially enormous domestic demand in mainland Europe. However, because of the complexities of the movement of spices in bulk, their price could increase almost twelvefold due to the duties placed on their transportation as they moved from the islands of the Indonesian archipelago to the marts of northern Europe. Behaim was clearly aware that the establishment of a seaborne passage to the vast trading emporia of Southeast Asia would give the Portuguese unique access to the trade in spices, thus circumventing the lengthy and costly overland trade routes through the Middle East and the Mediterranean which had previously brought them to Venice at such vastly inflated prices. The globe therefore operated as both a statement of intent on behalf of the rapidly moving Portuguese voyages, and an incentive to German financiers to invest in the commercial advances of the Portuguese which, it was hoped, would make the movement of spices through Venice a thing of the past.

At the same time as Behaim's globe was being completed, news began to circulate throughout the courts of mainland Europe of Christopher Columbus' discovery of land in the western Atlantic. Columbus' return to Europe in the spring of 1493 sparked a diplomatic controversy between the crowns of Portugal and Castile as he claimed the newly discovered territories on behalf of his sponsors, the Castilian crown, much to the consternation of João II. Like Columbus, João believed that the Genoese navigator had managed to discover a seaborne route to the spice markets of the east, by sailing westwards. Barros records João's anger at his return with evidence which seemed to suggest that he had in fact discovered the route to India. The king:

... was very upset to see that the people who came with him [Columbus] did not have black, kinky hair like those of Guinea, but rather resembling the colour of hair which he was informed was like that of India over which he had labored so hard.[27]

João contested Columbus' discoveries as an infringement upon territories that the Portuguese had already claimed under the Treaty of Alcáçovas, which had been signed between the crowns of Portugal and Castile in

1479. This had stipulated that Portuguese influence extended to any territories discovered 'beyond' Guinea, but quite how far this 'beyond' stretched was tested to its limits by Columbus' discoveries. João sought papal arbitration in his dispute with Castile. After extensive negotiation between the crowns and Pope Alexander VI a compromise was reached in June 1494 under the terms of the Treaty of Tordesillas. Castile claimed the territories discovered by Columbus, whilst the Portuguese crown was ensured exclusive rights to all navigational routes throughout Africa, the eastern Atlantic and all territories to the east. The final Treaty of Tordesillas stipulated that:

> . . . a boundary or straight line be determined and drawn north and south, from pole to pole, on the said ocean sea, from the Arctic to the Antarctic pole. This boundary or line shall be drawn straight, at a distance of three hundred and seventy leagues west of the Cape Verde Islands.[28]

All territory to the west of this line fell under the jurisdiction of Castile. All territory to the east was allocated to Portugal. Subsequent maps, like the Cantino Planisphere produced in 1502, dutifully inscribed the dividing line between the two crowns, not only in an attempt to demarcate the relative spheres of interest of Portugal and Castile, but also to establish a geographically agreed meridian from which to define the significance of further discoveries.

As a result of Columbus' discoveries and the outcome of the Treaty of Tordesillas, it became increasingly imperative for the Portuguese to push on with their search for a navigable route to India before Castile could dispatch new voyages to further erode their claims to the east. The death of João in 1495 led to a brief delay in further seaborne initiatives, but in 1497 the new king, Manuel, ordered the Portuguese commander Vasco da Gama to embark on a voyage which was to capitalize on Dias' earlier voyage round the Cape of Good Hope by reaching India and establishing trading relations with the spice markets of the region. Embarking in July, da Gama's fleet sailed out into the Atlantic and headed south. After rounding the Cape the fleet sailed up the coast of East Africa as far as Malindi, where they struck out across the Indian Ocean. By May 1498 da Gama's ships had landed in Calicut on the west coast of India, and the Portuguese immediately began to negotiate trading agreements with the local ruler, the Samorin of Calicut. The tentativeness of da Gama's encounter with the samorin and his court is emphasized in the bizarre array of gifts presented to the king:

> . . . twelve pieces of *lambel* [striped cloth], four scarlet hoods, six hats, four strings of coral, a case containing six wash-hand basins, a case of sugar, two casks of oil, and two of honey.[29]

The Muslim factors operating out of Calicut inspected the motley offering and poured scorn on the Portuguese gifts:

> They came, and when they saw the present they laughed at it, saying it was not a thing to offer to a king, that the poorest merchant from Mecca, or from any other part of India, gave more, and that if he wanted to make a present it should be in gold, as the king would not accept such things.[30]

Not surprisingly, the samorin was suitably unimpressed by the gifts, and after tense negotiations the Portuguese engaged in limited bartering with local merchants. By August, as relations with the samorin deteriorated, da Gama set sail on the return journey. In September 1499 he arrived back in Lisbon with a small but precious cargo of cinnamon, cloves, ginger, nutmeg and pepper, as well as a variety of woods and precious stones, enough to cover the expenses of the voyage six times over.

The news of da Gama's successful expedition was received with as much excitement as that accorded to Columbus. Its reverberations were quickly felt throughout mainland Europe as well as on the Iberian peninsula. Da Gama had opened the way, albeit tentatively, for Portuguese participation in the potentially lucrative trade of the Indian Ocean. King Manuel immediately wrote to the Castilian crown, triumphantly announcing the impact which da Gama's voyage would have on the commerce of early modern Europe:

> Moreover, we hope, with the help of God, that the great trade which now enriches the Moors of those parts, through whose hands it passes without the intervention of other persons or people, shall, in consequence of our regulations be diverted to the nations and ships of our own kingdom, so henceforth all Christendom, in this part of Europe, shall be able, in a large measure, to provide itself with these spices and precious stones.[31]

Manuel's rhetoric of Christian unity and expansion in the face of the power of Islam neatly forestalled any potential Castilian objections to Portugal's claim to these newly established trade routes via the Cape of Good Hope.

However, the Venetian state was not as easily appeased as Castile, and in 1500 an alarmed Venetian seigniory dispatched a diplomatic delegation to Lisbon, led by Pietro Pasqualigo. At the level of formal exchanges the Venetians conceded the commercial and political significance of the Portuguese achievements. In his diplomatic address delivered to Manuel, Pasqualigo announced in his discussion of the newly established trade route:

> What is greatest and most memorable of all, you have brought together under your command peoples whom nature divides, and with your commerce you have joined two different worlds.[32]

With a similar rhetorical flourish Pasqualigo went on to claim:

It should be understood how much benefit and utility the whole world gains from the cane sugar and the many other products which are then transported in great abundance in all directions for the use of mankind, and how much even posterity will benefit from the spices of every sort recently found by your ships.[33]

However, behind the scenes the Venetians desperately tried to sabotage Portugal's increasingly assertive participation in the spice trade. According to Barros, Pasqualigo's entourage even infiltrated a diplomatic delegation from Indian states trading in spices, who had arrived in Lisbon to discuss Portugal's growing interest in the trade. The Venetians tried to argue that the Indians would benefit more by dealing directly with Venice rather than Portugal.[34] The fears of the Venetian delegation were well founded, and were echoed in the gloomy predictions made by prominent merchants in Venice on hearing the news of Portugal's arrival in India. Girolamo Priuli, a merchant and diarist, gloomily recorded in his diary in 1502:

Therefore, now that this new route has been found by Portugal this King of Portugal will bring all the spices to Lisbon and there is no doubt that the Hungarians, the Flemish and the French, and all the people from across the mountains who once came to Venice to buy spices with their money will now turn to Lisbon because it is nearer to their countries and easier to reach; also because they will be able to buy at a cheaper price, which is most important of all. This is because the spices that come to Venice pass through all of Syria, and through the entire country of the Sultan and everywhere they pay the most burdensome duties. Like wise, in the state of Venice they pay insupportable duties, customs, and excises. Thus with all the duties, customs, and excises between the country of the Sultan and the city of Venice I might say that a thing that cost one ducat multiplies to sixty and perhaps to a hundred.[35]

In terms strikingly similar to the notes made by Behaim on his terrestrial globe, Priuli quickly grasped the significance of the establishment of a direct seaborne trade route with India which circumvented the need to pay the expensive customs duties for which the Venetian merchants found themselves liable. He pessimistically concluded that:

. . . if this voyage from Lisbon to Calicut continues as it has begun, there will be a shortage of spices for the Venetian galleys, and their merchants will be like a baby without milk and nourishment. And in this, I clearly see the ruin of the city of Venice.[36]

As successive Portuguese voyages to India strengthened Manuel's grip on the importation of spices into mainland Europe, Priuli's predictions appeared at first to be only too accurate, a situation which led to an intensification of the diplomatic conflict between Portugal and Venice. Underlining the illusion of Christendom uniting in the face of the growing power of the Islamic Ottoman empire, Venice urged both the Ottoman sultan Bayezid II and the Egyptian Mamluk sultans to unite in a three-pronged diplomatic and, if necessary, military defence of their common trading interests. In 1503 Venice signed a peace agreement with the Ottomans in an attempt to stabilize trading relations in the eastern Mediterranean. By 1511 the Venetian seigniory had dispatched diplomatic delegations to both Istanbul and Cairo in an attempt to encourage military actions against the commercial incursions of the Portuguese in India.

Nor did the Portuguese fly the unblemished flag of crusading Christianity regardless of their own commercial interests. In 1511 the naval commander Alfonso d'Albuquerque negotiated with the Persian ruler Shah Ismail to launch a joint military attack on Egypt in an attempt to secure strategic points in the Red Sea, which would ensure Portugal's expanding commercial development and divest Venice of one of its most lucrative trade routes to the east.[37] Whilst the plan ultimately failed to materialize, it was symptomatic of the extent to which the Portuguese discovery of the seaborne route to the markets of the Indian Ocean brought a whole new dimension to the commercial, diplomatic and geographical shape of the early modern world. Portugal's emergence as one of the most powerful empires in early sixteenth-century Europe was coterminous with its emergence as a key player in the commercial and political world of the Indian Ocean and Southeast Asia. The political suspicion and commercial jealousy directed towards the Portuguese, not only by Venice but also by the crowns of France and England, was symptomatic of the anxiety produced by Portugal's ability to generate its conspicuous wealth and power from a strategic network of commercial and political locations which were far beyond the reach of mainland Europe, and which were built, very literally, on water.

This remarkable dispersion and diversification of long-distance travel, commercial exchange and diplomatic negotiation produced by the consequences of Portugal's establishment of the Cape route to India required logistical and imaginative apparatuses to make sense of such changes. Maps, charts and globes such as the Cantino Planisphere and Behaim's terrestrial globe disseminated vital conceptual information on the changing territorial and commercial shape of the world they depicted. Supplemented by the mass of travel narratives and diplomatic reports which emerged as a result of the Portuguese activities, such geographical artefacts became prized possessions, not only keeping their owners informed of the latest discoveries and commercial ventures, but also providing them with a sense of security as to their own identity within such an ever-changing world. To be aware of the changing nature of the world was to be able to position oneself confidently in relation to that world, even if the news that came from it may not always have been welcome. In August 1501, at the beginning of the diplomatic conflict between Venice and Portugal, Angelo Trevisan, secretary to the Venetian ambassador in Spain, wrote to the Venetian annalist Domenico Malipiero from Granada, announcing that:

We are daily expecting our doctor from Lisbon, who left our magnificent ambassador there; who at my request has written a short account of the voyage from Calicut, of which I will make a copy for Your Magnificence. It is impossible to procure the map of that voyage because the king has placed a death penalty on any one who gives it out.[38]

However, by the end of September Trevisan wrote again, informing Malipiero that:

If we return to Venice alive, Your Magnificence will see maps both as far as Calicut and beyond there less than twice the distance from here to Flanders. I promise you that everything has come in good order; but this, Your Magnificence may not care to divulge. One thing is certain, that you will learn upon our arrival as many particulars as though you had been at Calicut and farther.[39]

It is not clear which map Trevisan obtained for Malipiero, but it presumably drew on the appearance of contemporaneous maps like the Cantino Planisphere, or the subsequent world chart attributed to Nicolaus de Caverio (illus. 25), which in its extensive depiction of the African coastline and extensive legends on the trade at Calicut and Cochin reflected detailed awareness of the most recent Portuguese voyages. The desire for precise information on the Portuguese activities, whatever their implications, created a new climate for the traffic in maps and charts as valuable material objects. In a diplomatic context, where the failure to obtain accurate information on the activities of a rival crown in distant places spelt potential political and commercial disaster, the strategic retention or dispersal of geographical information took on added significance. By 1510 such information was being offered for distribution on the open market. The Lisbon-based German printer Valentim Fernandes, who was responsible for printing the *Reportório dos Tempos* in 1518, wrote to his colleague Steffan Gabler in 1510:

I want to send you the coast from India to Malacca and the map with the islands, because so far all the pilots are still busy in the king's house; but afterwards they are also at my disposition.[40]

As with the Cantino Planisphere, it becomes clear that the latest geographical information contained in the maritime charts of the Portuguese was available to the merchants and diplomats of mainland Europe. However, like the spices and pepper which began to appear in the marts of northern Europe, this was at a price fixed within an intellectual market-place that was often as competitive as the marts which bought and sold the precious spices from India.

The 'Continuous Travailes' of Portuguese Cartography

If one of the reasons why the merchants and diplomats of the early modern world valued the maps and charts of the early Portuguese discoveries was for their ability to map and subsequently expand the margins of the known world, it is ironic but significant that this very process should come to constitute one of the central reasons for the intensified social and cultural ambivalence accorded to the Portuguese over subsequent centuries. It is in the nature of élite travellers like merchants and diplomats to explore and explain the margins of their world picture. For merchants in particular, and commercially minded monarchs like Dom Henrique, a gap on the edge of the map could invariably constitute a gap in the commercial market. However, such investigations of the cultural and geographical margins has historically always evoked fear and suspicion in many people. It is this ambivalence which is central to repeated accounts of the Portuguese as exorbitant voyagers, heroically sailing beyond the limits of their own culture but risking exposure to any number of dangerous and debilitating situations, from death and disease to more mysterious and intangible dangers associated with figures who leave their own culture and subsequently return from places unknown to the majority of the population. Throughout the period writers warned of the dangers of travel to unknown territories, and the potentially damaging effect it could have on the traveller. Yet maps and charts were problematically positioned in relation to this ambivalent response. Whilst maps and charts may assist the armchair geographer in avoiding the need to travel, such geographical information must inevitably emerge from direct encounters and transactions with distant places and people. One convenient solution to this paradox was to elide the contexts of the production of such maps. In the Preface to his world map published in 1506, the Venetian cartographer Giovanni Contarini announced boldly:

The world and all its seas on a flat map . . .
Lo! Giovanni Matteo Contarini,
Famed in the Ptolemaean art, has compiled and marked it out.
Whither away? Stay, traveller, and behold new nations and a new-found world.[41]

Self-consciously selling his new map on the basis of its ability to insulate its buyer from the potentially dangerous effects of long-distance travel, Contarini's comment neatly sidesteps the fact that it was produced from the maritime charts of the early Portuguese voyages along the African coast and into the Indian Ocean. By 1559 the English geographer William

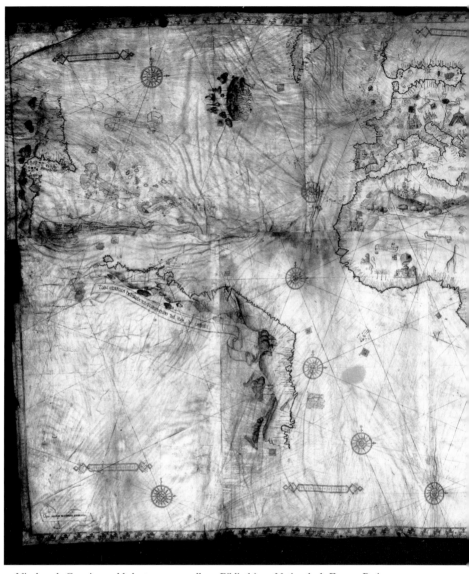

25 Nicolaus de Caverio, world chart, *c.* 1505, vellum. Bibliothèque Nationale de France, Paris (Charts and Plans).

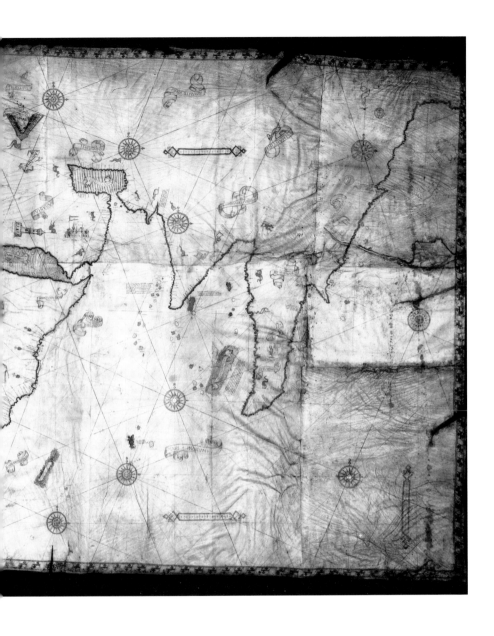

Cunningham could write of his experience of drawing up maps that:

I may in like sorte at my pleasure, drawe a Carde for Spaine, Fraunce, Germany, Italye, Graece, or any perticuler region: yea, in a warme & pleasant house, without any perill of the raging Seas: danger of enemies: losse of time: spending of substaunce: werines of body, or anguishe of minde. O how precious a Jewell is this, it may rightly be called a Cosmographicall Glasse, in which we may beholde the diversitie of countries: natures of people, & innumerable formes of Beastes, Foules, Fishes, Trees, Frutes, Stremes, & Meatalles.[42]

For Cunningham, the advantages of such maps and charts was that they 'delivereth us from greate and continuall travailes', exemplifying a contemporary conflation of the idea of travel as inevitably involving painful exertion and personal suffering. If his position became one of the selling points of lavishly printed sixteenth-century maps and atlases, then it was in part as a result of the practical advances made by Portuguese pilots, merchants and sailors as they inched their way along the coast of Africa and carefully pinpointed their journeys onwards towards India. The Portuguese did not have the luxury of taking up such a position, and in many ways were subsequently judged for their inability to retain a level of cultural 'distance' from the communities with which they interacted. Their sustained encounters and transactions with peoples and cultures perceived as either barbarous or antithetical to the principles of European civility marked them as outsiders within the conception of Europe itself, dangerous voyagers who persistently exposed themselves to the deracinating effects of long-distance travel.

Whilst the markets of mainland Europe may have been eager to consume the precious merchandise loaded by the Portuguese ships in Guinea, Calicut, Goa and Malacca, its scholars were more circumspect in their acceptance of knowledges gleaned from communities which it deemed inimical to the values of European culture. Yet the Portuguese were to stretch the intellectual and geographical limits of European knowledge in their voyages and explorations, leading them to encounters and transactions with peoples and cultures who, as in the case of da Gama's landing in India, regarded their learning and aspirations as at best incomprehensible and at worst highly inadequate. However, the pragmatic Portuguese persistently assimilated the alien knowledges which they encountered and fed them straight into the maps and charts which they gradually built up as they sailed further into unknown waters. In 1498, as Vasco da Gama sailed up the east coast of Africa, he quickly realized how inadequate his maps and navigational instruments were in attempting to cross the Indian Ocean to reach India. Whilst stopping in Malindi in 1498, his crew employed a Muslim navigator. In the written account of the voyage, the narrator explains that whilst in Malindi:

... there came a Moor, a Gujerati named Malemo Caná who, as much because of the pleasure which he had from conversation with us, as in order to please the King [of Malindi] who was seeking a navigator for him [da Gama], agreed to go with him. Vasco da Gama, after he had had a discussion with him, was greatly satisfied with his knowledge: principally, when he [the Moor] showed him a chart of the whole of the coast of India drawn, in the fashion of the Moors, that is with meridians and parallels, very close together but without any wind rhumbs. Because the square network of these meridians and parallels was very small, the coast within those from North to South and from East to West was very accurate, without that multiplication of wind rhumbs drawn from a single compass-rose of our charts, which serves as the origin of the other directions. And when Vasco da Gama showed him a large astrolabe of wood which he had with him, and others of metal with which he measured the altitude of the sun, the Moor expressed no surprise, saying that some navigators of the Red Sea used brass instruments of triangular shape and quadrants with which they measured the altitude of the sun and, principally of the star [the Pole Star] which they most commonly used in navigation.[43]

Caná offered da Gama hydrographical information which utilized a much more sophisticated knowledge of astronomical observation than had been developed even in Portugal. Eschewing the techniques practised in the Mediterranean of navigating according to the compass and wind direction, Caná's charts and knowledge of navigational instruments were crucial to the successful navigation from Malindi to Calicut, and da Gama clearly had no scruples in utilizing a Muslim pilot in the pursuit of his mission. In the triumphant letter which King Manuel sent to the Castilian crown in the aftermath of da Gama's voyage, the Portuguese sovereign displayed little reticence in celebrating the local navigational knowledges which da Gama appropriated whilst on his voyage:

... they captured a ship of the King of Calicut, from which they have brought to me some jewels of great price, 1500 pearls, amounting to 8000 ducats, three astrological instruments of silver, not in use amongst our astrologers. They are large and well made, and have been extremely useful to me. They say that the King of Calicut had sent the aforementioned ship to an island called Saponin to obtain these instruments and to get a good pilot and a navigation chart for those lands. Now the pilot is in our hands, and I am having our language taught to him since he shows an understanding of these astrological instruments.[44]

The information gleaned from local pilots and cosmographers appears to have been rapidly assimilated into the early Portuguese charts of the seas and territories to the east of the Cape. Whilst negotiating with the Persian Shah Ismail in 1511, Alfonso d'Albuquerque obtained a detailed chart depicting the Cape of Good Hope, the Red Sea, the Indian Ocean and the Moluccas. It was constructed by a Javan pilot, and the place names were written in Javan. A portion of the chart was forwarded to Manuel by the pilot, Francisco Rodrigues, who appended a legend to the map, announcing:

Your Highness can truly see where the Chinese and Gores come from, and the course your ships must take to the Clove Islands, and where the gold mines lie, and the islands of Java and Banda, of nutmeg and maces, and the land of the King of Siam, and also the end of the navigation of the Chinese, the direction it takes, and how they do not navigate further.[45]

Rodrigues had sailed to Malacca with d'Albuquerque in 1511 and went under the grand title of 'Pilot-major of the first armada that discovered Banda and the Moluccas'. He was reputed to be a Javan seconded by d'Albuquerque whilst on his travels, and in 1513 he produced a manuscript book of *roteiros*, or maritime charts, supplemented with written navigational descriptions of coasts. These were amongst the first charts of southern and eastern Africa to calculate the latitudes of the region (illus. 26), presumably by using the types of astronomical instruments Caná had told da Gama about. Maps like the Cantino Planisphere and the world chart of Caverio both show evidence of their absorption of local knowledges in their calculation of latitude and depiction of territory.[46] Both are indicative of the extent to which the maps and charts which emerged from these Portuguese voyages were not the pristine and triumphant cultural trophies of Renaissance exploration, but hybrid cultural artefacts fashioned from the exchange of knowledge and information between the Portuguese and local interlocutors. The elaborately printed global geography produced in the print capitals of northern Europe, such as Martin Waldseemüller's world map of 1507 (illus. 27), whilst drawing extensively on the Portuguese voyages and their hybrid maps and charts, expunged all reference to the ways in which their production was informed by the 'alien intelligence' gleaned from local pilots and geographers. Whilst Waldseemüller's map attempted to incorporate the Castilian discoveries in the Americas to the west, it is clear that it is the territories to the south and east, namely Africa and Asia, which still held the main focus of geographical attention. It was magnificently printed maps like Waldseemüller's that captured the attention of the map-buying élite of early modern Europe, rather than the scarce and precious manuscript maps and charts produced by Portuguese pilots and geographers. As a result, such maps carefully occluded the alien knowledges assimilated into the original Portuguese charts, offering their owners a sanitized vision of geographical and commercial expansion, bereft of the troublingly unfamiliar encounters which the Portuguese experienced on their long-distance voyages.

The fact that the Portuguese built their extensive commercial empire

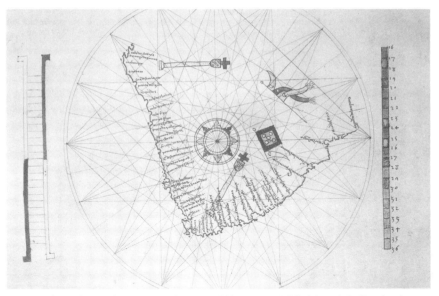

26 Francisco Rodrigues, chart of southern Africa, *c*. 1513. Bibliothèque de la Chambre des Deputés, Paris.

literally on water only fuelled suspicions as to their political power and cultural authority. Of all forms of long-distance travel none was perceived with more fear and awe than travelling by sea. Voyaging across unknown oceans to relatively unknown lands only reinforced people's suspicions as to where the Portuguese had gone, and from where they would return. Seaborne travel also brought with it practical difficulties which forestalled any possibility of establishing recognizably systematic regimes of colonial domination for centuries to come. Transporting men and equipment by sea over long distances was a necessary evil, as it was invariably more commodious than travelling overland. However, it was a slow, laborious and dangerous undertaking. Unlike the relentless territorial expansion of the Castilian crown in the Americas, which concentrated on plunder and the extraction of tribute, the Portuguese empire was always much more reliant upon a far more transactional and commercial model of development. As a result, it had to come to terms with knowledges which it encountered which could not be simply ignored or destroyed, as they were vital to the commercial success of the Portuguese initiatives. The production of maps and charts exemplified the transactional nature of the early Portuguese voyages down the coast of Africa and onwards into the Indian Ocean. They became, in fact, the practical bedrock for all subsequent printed cartography over much of the globe for the next 200 years, however distasteful mainstream European reactions to the Portuguese intermingling

27 Martin Waldseemüller, world map, woodcut on 12 sheets of paper, Strassburg, 1507.
Fürstlich zu Waldburg Wolfegg'sche Kupferstichkabinett.

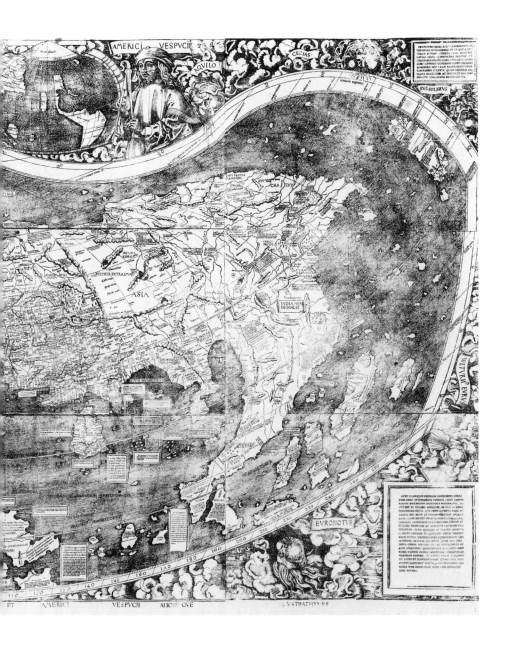

of goods, people and knowledge became. Perhaps now, bereft of a perception of the Renaissance as a time of unrivalled European confidence in the pursuit of new worlds in the name of an unbounded curiosity, and in the light of revisions within Portuguese historiography itself as to the status of the 'Great Discoveries',[47] it is possible to reassess the importance of the Portuguese who, in their pursuit of worldly wealth, were primarily responsible for defining the political and commercial contours of the early modern world.

3 Disorienting the East: The Geography of the Ottoman Empire

Plotting with the Enemy?

In 1464 the Florentine scholar Francesco Berlinghieri began work on a geographical treatise based on Ptolemy's *Geographia*, updating its maps and supplementing them with a running commentary composed in verse. Berlinghieri was an established figure in the humanist community which had begun to emerge in late fifteenth-century Florence. After finding employment as an orator at the Este court in Mantua, he subsequently joined the court of Lorenzo de' Medici and participated in the Platonic Academy established in Florence by Marsilio Ficino, under the patronage of Lorenzo. Berlinghieri was amongst the *auditores* of Ficino's lectures which extolled the values of classical Greek learning.[1] He also took part in a public disputation with Ficino on the intellectual significance of Ptolemy's *Geographia*. Berlinghieri was therefore uniquely situated at the centre of humanist learning in late fifteenth-century Italy, a situation reflected in the publication of his geographical treatise based on Ptolemy's highly influential geographical text. By the summer of 1482 the translation was finally complete and ready to be printed, with maps, by the renowned German printer Nicolaus Laurentii, under the title *Septe Giornate della Geographia di Francesco Berlinghieri*.

   It is difficult to overestimate the significance of Berlinghieri's printed treatise in the early history of cartography. Not only was it amongst the first texts based on Ptolemy's *Geographia* to be actually printed, but it was also the first work on Ptolemy to be translated into vernacular Italian. Perhaps even more importantly, Berlinghieri's text was one of the first to break with Ptolemaic tradition by supplementing the accepted running order of the maps in the *Geographia* with four 'modern' regional maps of Italy, Spain, France and Palestine.[2] As a result it established a counter-tradition which involved the gradual incorporation of new maps and geographical information into subsequent printed collections of maps which based themselves on Ptolemy. Although the world map which adorned the first pages of the section of maps in Berlinghieri's *Geographia* remained tied to the Ptolemaic conception of the *oikoumene* (illus. 28), its utilization of the new techniques of print culture produced an image of remarkable

28 Francesco Berlinghieri, world map from his *Geographia*, copper engraving, Florence, 1482. British Library, London.

visual clarity and geographical precision, which surpassed previous manuscript attempts to re-create the Ptolemaic *oikoumene*. Like Berlinghieri
himself, the elegant and erudite text stood as a striking example of the
new humanist learning which was spreading across the city-states of late
fifteenth-century Italy, as scholars, painters and architects turned to the
classical worlds of Greece and Rome for inspiration and instruction
within the fields of science, education and the arts. However, all was not
as it seemed, for Berlinghieri's innovative text was not, as one might have
expected, dedicated to one of the many princely patrons throughout fifteenth-century Italy, such as Lorenzo de' Medici. Instead, Berlinghieri
chose to dedicate his treatise to none other than the Ottoman sultan
Mehmed II, 'the Conqueror', the so-called 'scourge of Christendom', who
in 1453 had captured the ancient capital of the Byzantine empire, Constantinople, from the disparate military forces of Christian Europe. Even
more remarkable is the surviving record of Berlinghieri's admiring dedication to Mehmed in the copy of the text which still exists in the Topkapi
Saray Museum in Istanbul:

To Mehmed of the Ottomans, illustrious prince and lord of the throne of God,
emperor and merciful lord of all Asia and Greece, I dedicate this work.[3]

Standard histories of the European Renaissance have invariably defined
the development of its love of learning and commitment to a civilizing
process in direct opposition to the menacing threat of the Islamic Ottoman
empire, persistently represented as a dark despotic threat looming over the
forces of enlightenment and disinterested intellectual enquiry which supposedly characterized late fifteenth- and early sixteenth-century Europe.
Within this scenario, the Ottoman empire, and Mehmed in particular,
came to represent the direct antithesis of everything for which the European Renaissance stood. However, if this was indeed the case, then how
did such an intellectually and politically respected figure as Berlinghieri
come to dedicate one of the most innovative geographical texts of the
Italian Renaissance to one of its most apparently demonized figures?
This chapter seeks to provide an answer to this question by rejecting
accounts of the Renaissance which continue to demonize and occlude the
place of the Ottoman empire within the social and cultural dynamics of
the early modern period. To place the Ottomans beyond the remit of the
development of early modern Europe is to emplot a much later geopolitical shift upon an earlier historical moment, a shift which fixed the
boundary between western Europe and the Ottoman territories at a point
of military disengagement at the beginning of the eighteenth century.[4]
The concern of this chapter is to examine what the cultural and geographical landscape of early modern Europe looked like prior to the imposition

of these frontiers between East and West, and to what extent the Ottoman empire was constitutive of, rather than antagonistic to, the formation of the cultural and geographical map of late fifteenth- and early sixteenth-century Europe. The following pages therefore attempt to reorient an understanding of early modern Europe and its boundaries which places the Ottomans as central, rather than peripheral, to the political and intellectual preoccupations of the period.

By the middle of the sixteenth century, the Ottoman empire was one of the most powerful empires in the early modern world. Its borders stretched for over 10,000 kilometres, encompassing huge swathes of territory from Hungary to the steppes of Central Asia, and it outstripped the realms under the control of any comparable European monarch.[5] Its origins emanated from a small Anatolian province located between Byzantium and the Seljuk Turks, which grew throughout the fourteenth century until, by the fifteenth century, the rapidly growing Muslim forces of the Ottomans threatened the seat of the Holy Roman Empire itself, Constantinople. In 1453 Constantinople (hereafter referred to as Istanbul) finally fell to Mehmed's forces. Standard western critical responses to the fall of the city have characterized the event as one of the most disastrous calamities to befall Christian Europe in the fifteenth century, inspiring fear and dread in the hearts of Christians throughout Europe that the crescent flag of Islam would ultimately overrun the length and breadth of mainland Europe.[6] However, the capture of the city by the Ottomans and their subsequent annexation of the territories of the former Byzantine empire left the scholars, merchants and diplomats of fifteenth-century Italy exhibiting far more ambivalent responses to the rise of Ottoman power than has often been believed. Within just one year of the fall of the Byzantine capital, the Venetian authorities had signed a peace treaty with the Ottoman authorities that allowed them to continue trading in regions previously under the control of the Byzantine empire, which now came under Ottoman jurisdiction. Scholars of the period were also far more circumspect in their responses to the impact of the Ottoman presence in Istanbul. The territories which increasingly came under the control of the Ottomans were precisely those areas from which early Italian humanists like Ficino and Berlinghieri had drawn their intellectual and imaginative sustenance in their recovery of the learning of classical antiquity. Stories abound of the desperate transmission of precious Greek and Roman manuscripts between the Byzantine city and mainland Italy prior to the fall of the city. However, scholars appear to have been more concerned that, rather than rejecting the importance of classical learning, the Ottomans were actually appropriating classical Graeco-Roman models of learning for their own political ends. In 1466 the Greek scholar George Trepuzuntios wrote to Mehmed the Conqueror informing him that:

No one doubts that you are emperor of the Romans. Whoever holds by right the center of the Empire is emperor and the center of the Empire is Constantinople.[7]

Another Greek scholar, Kritovoulos of Imbros, described Mehmed's achievements as being 'in no way inferior to those of Alexander the Macedonian'.[8] In 1481 the Venetian painter and medallist Costanzo da Ferrara arrived in Istanbul at the request of Mehmed, and cast a portrait medal of the Ottoman sultan (illus. 29) which drew extensively on Roman imperial prototypes, a fact which was presumably not lost on the astute Mehmed. In terms of the newly acquired geographical position of the seat of Ottoman power, Mehmed's reign sought to cultivate political and cultural comparisons and analogies between the Graeco-Roman classical tradition and the successes of the Ottoman empire, rather than establishing an attitude of hostility or incomprehension towards such a tradition, which would only have been counterproductive in making Mehmed's cultural, and hence political, power intelligible to his Italian political counterparts. Such comparisons were also not lost on the rival city-states of Italy as they jockeyed for authority towards the end of the fifteenth century. The Ottomans appeared to many as highly desirable political allies in the delicate balance of power within the eastern Mediterranean. In 1461 Sigismondo Malatesta, Lord of Rimini, sent the painter, architect and medallist Matteo de' Pasti to the Ottoman court, with a letter praising Mehmed's military exploits and, like Kritovoulos, comparing his deeds to those of Alexander the Great.[9] Malatesta's reasons for courting the sultan so assiduously seem to have been based on his search for a political ally in his continuing rivalry with Venice. The Ottomans did not therefore cut off political and cultural connections with the Graeco-Roman world which mainland Europe had been cultivating throughout the fifteenth century; they simply appropriated them as a way of strengthening their own claims to imperial authority throughout the territories of the classical world. It was this fact which concerned the scholars and diplomats of early modern Europe. Whilst the public diplomatic rhetoric of the period condemned the expansion of the Ottomans and called for a united Christian front against the growth of a powerful Muslim empire, behind the scenes city-states like Venice, Rimini and Florence fought for political and commercial concessions from the ascendant Ottoman empire.

Similarly, scholars like Ficino and Berlinghieri were keen to retain links with Istanbul as one of the remaining centres of learning which established a connection between the classical world of ancient Rome and Greece and

29 Costanzo da Ferrara, *Sultan Mehmed II* (obverse), *The Sultan Riding* (reverse), *c.* 1481, bronze portrait medal. National Gallery of Art, Washington, DC.

the contemporary world of the fifteenth century. As the Ottomans controlled a significant area of the *oikoumene* portrayed in Ptolemy's *Geographia*, they were also in possession of not only the majority of the most lucrative overland trade routes between Asia and mainland Europe, but also most of the ancient centres of learning which were celebrated by fifteenth-century Italian humanism, such as Baghdad, Alexandria and Cairo. As with the Graeco-Roman world, this did not involve Ottoman imperial ignorance or dismissal of the intellectual traditions represented by these centres of learning, but an appropriation of such traditions for their own ends. As the Ottomans were so centrally involved in both the diplomatic and intellectual worlds of late fifteenth- and early sixteenth-century Europe, it is telling that a prominent scholar like Berlinghieri should choose to dedicate such a significant geographical text as his *Geographia* to an illustrious patron of the calibre of the Ottoman sultan Mehmed II. Quite how central the Ottomans were in the diplomatic and intellectual controversy which surrounded the production of Berlinghieri's treatise becomes clear if we take a closer look at its fascinating history.

Berlinghieri's text stands in its own right as a fine example of the practical and intellectual issues which emerged around the new technology of printing. Berlinghieri's *Geographia* was over thirteen years in the making, as the Florentine scholar collated the works of Ptolemy as well as supplementing them with more contemporary maps and geographical treatises. As he calculated that the text would include thirty-one maps (more than any other printed text then available), it was decided that the illustrations should be engraved on copper plates before being printed. This was an expensive and highly sophisticated printing technique which allowed for higher technical definition in illustrations than that produced using woodcuts. However, as there were few established copper-engravers skilled in the art of cartography, Berlinghieri had to wait longer than usual for the completion of his maps. As initial copies of his text slowly emerged from the printing house, Berlinghieri took the initiative of ascribing manuscript dedications in the front of selected printed copies. One such copy contained the manuscript dedication to Mehmed II. In this copy of his *Geographia* Berlinghieri inserted a letter dedicating the text to the new sultan, Bayezid II. In the letter he explained that he had intended to dedicate his *Geographia* to Bayezid's father, Mehmed, but that when the news reached Florence that Mehmed had died in May 1481 he had been left without a figure of suitable stature to whom he could dedicate his text. As a result, he had hastily decided to dedicate it to the Duke of Urbino, Federigo da Montefeltro, whose successful career as a mercenary soldier was matched by his enthusiastic patronage of scholarship and the arts.[10] By 1482 Federigo's library contained over a thousand manuscripts, whilst contemporary paintings depicted him as a master of both the pen and the sword (illus. 30). Berlinghieri described the duke in appropriately glowing terms, lauding him as being 'supreme in military virtue and wisdom'. Unfortunately for the luckless Berlinghieri, by the time the corrected type for the edition had been finally completed Federigo had also died, in September 1482. Berlinghieri explained that he was therefore returning to his initial attempt to dedicate the text to an Ottoman sultan, and was dedicating it to Bayezid II in the light of the deaths of both Mehmed and Federigo. Whilst not as enthusiastic a patron of the arts as his father, Bayezid oversaw a range of intellectual and cultural transactions between Istanbul and the Italian city-states which had been encouraged by his father, including negotiations in 1492 with Francesco Gonzaga II of Mantua, who wished to buy Arab horses from the sultan.[11] Bayezid also established a reputation amongst the Florentine merchants based in Istanbul as a respected sovereign with whom they could do business. In 1502 the Florentine merchant Giovanni di Francesco Maringhi wrote to colleagues in Florence from Istanbul claiming that Bayezid:

. . . has shown so great an interest in us . . . [and] in all of our business; and since he also displays a cordial fondness for all our nation. Before I close this, I will tell you the kindness he has done, and is still constantly doing with respect to the dead and the bankrupt merchants; so that one is able to see that he desires that we be held in esteem throughout all his realm. May God send his blessing.[12]

News of the esteem in which Bayezid was held by the Florentine community since his assumption of royal power in 1481 can have only strengthened Berlinghieri's resolve to dedicate his text to the new Ottoman sultan. The fact that Berlinghieri repeatedly attempted to dedicate the

30 Pedro Berruguete, *Federigo da Montefeltro with his son Guidobaldo*, 1477, oil on canvas. Palazzo Ducale, Urbino.

Geographia to an Ottoman sultan, over and above potential Italian patrons like the Medici family, is powerful testimony to the political weight and cultural authority which emanated from the Ottoman court, an authority which was felt even amongst the scholars of fifteenth-century Florence.

However, the connection between Berlinghieri's *Geographia* and the Ottoman court did not end there. Mehmed's death in 1481 sparked a vicious struggle for political succession between his sons Bayezid and Cem. After being defeated in battle in 1482, Cem fled to Europe where he moved from court to court seeking support for an anti-Bayezid coalition to attack Istanbul and install himself as sultan. Despite the interest of several leading princes and monarchs, nobody was prepared to risk the level of military and financial expenditure required to realistically take on the might of Bayezid's armies. Eager to keep his brother out of the Ottoman dominions, Bayezid bought off the Christian Knights Hospitallers of Rhodes and Pope Innocent VIII himself with precious relics from his father's extensive collection, in an attempt to ensure that Cem remained a virtual hostage in Europe.[13] However, in the summer of 1484 Paolo da

Colle, an agent working on behalf of Lorenzo de' Medici, and who had taken Berlinghieri's *Geographia* to Istanbul in 1482 for presentation to Bayezid, was dispatched to Savoy, where Cem was living. Da Colle had been sent to Cem by Berlinghieri, and carried with him a copy of Berlinghieri's *Geographia*. The text which he delivered to Cem was identical to the text he had presented to Bayezid eighteen months earlier, down to the dedicatory letters written by Berlinghieri in both copies. The only change was a hastily added dedicatory preface in manuscript addressed to Cem, extolling his princely attributes in almost exactly the same terms as those addressed to Bayezid and Federigo.[14]

Berlinghieri's gift was little help to Cem in his attempt to wrest power from his brother. In 1495, whilst gathering a military force to attack Bayezid, he died mysteriously. Many believe he was poisoned by those sympathetic to Bayezid. Nevertheless, Berlinghieri's repeated attempts to associate his *Geographia* with Mehmed and with his two sons as they struggled for control of the Ottoman crown is symptomatic of the extent to which geography played a central role in the diplomatic manoeuvrings of the period, as the key protagonists in the struggle for the regions depicted in his maps circulated on paper the territories which they sought to possess by strength of arms. As such, Berlinghieri's dedicatory strategies were also a highly skilful marketing strategy to establish his *Geographia* as the primary geographical text through which the central protagonists in the diplomatic drama of the period came to imagine the territories across which they moved.

In technical terms, Berlinghieri's text did not completely live up to expectations. In trying to come to grips with the new techniques of printing from copper-engraving the maps were littered with misspelt names and hastily corrected misattributions. In the ninth map of the atlas entitled *Tabula Sexta De Uropa* depicting territories between mainland Europe and the Ottoman dominions, the word 'Uropa' of the title is run into the word 'Asia' (illus. 31). This striking error in the composition of the printed text is an example of the often contradictory procedures by which the institution of printed geography attempted to define the boundaries of early modern Europe. Whilst the mistake is symptomatic of the difficulties which the geographer and printer experienced in their attempts to utilize the new technology of print (it is noticeable that a similar mistake is also made in the sixth map of Europe), it is also a significant comment on the dedicatory vicissitudes and diplomatic manoeuvrings which framed the production of Berlinghieri's text.[15] At the historical point at which the boundary between Europe and Asia was itself in flux as a result of the rapid expansion of the Ottoman empire, the maps in the *Geographia* exhibit a level of uncertainty as to where contemporary geographers drew the

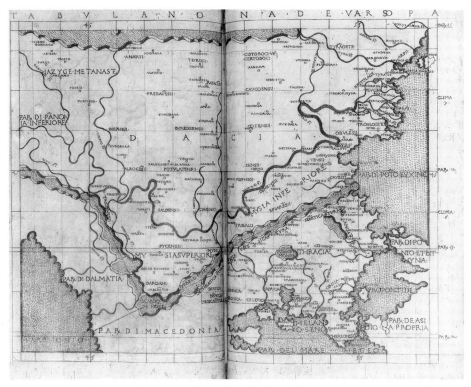

31 Francesco Berlinghieri, 'Tabula Sexta De Uropa', *Geographia*, copper engraving, Florence, 1482. British Library, London.

dividing line between 'Europe' and its eastern frontiers, which were under constant revision in the face of Ottoman territorial expansion. The princes and diplomats of the period consulted their personal maps in order to ascertain the status of their dominions as a result of territories traded between empires and states. The ease with which Berlinghieri negotiated the status of his own text in relation to the diplomatic and intellectual exchanges between the territories of the Ottoman empire and mainland Italy emphasizes the extent to which no such geographical or imaginative line of demarcation firmly existed between a political East and West in the early modern world. Whilst derogatory post-Enlightenment discourses of Orientalism have firmly demarcated the notion of the East in the western intellectual and political imagination as beginning somewhere around Istanbul, the geographical, diplomatic and intellectual exchanges across this apparent divide throughout the late fifteenth and early sixteenth centuries underline the fact that no such distinction existed for Berlinghieri and his contemporaries. The Ottomans were politically and intellectually powerful participants in the early modern world, and their leaders were as compliant and enthusiastic in the patronage of scholarship and artistic production as their Italian counterparts.

Mehmed's Maps: Ottoman Patronage of Geography

It should be stressed that Berlinghieri's attempts to dedicate his *Geographia* to Mehmed and Bayezid were not based on diplomatic expediency alone. In many ways even a scant knowledge of Mehmed's patronage of art and science would have convinced him that the sultan was the logical choice as patron of a geographical text based on the work of Ptolemy. Critics of the transmission of classical Graeco-Roman texts between Constantinople and Italy prior to the fall of the Byzantine capital have often portrayed this movement of learning and ideas as a seamless process whereby the scholars of fifteenth-century Italy imbibed the texts of the ancient world by a process of intellectual osmosis. Such a transmission of ideas has been seen as the backbone to the establishment of the so-called civilization of Renaissance Europe. The scholars and book-hunters of the period did systematically track down ancient manuscripts and methodically engage in the massive task of translating these texts from the original Greek into Latin for wider consumption. However, the status of Arabic learning and, even more noticeably, the role of the Ottomans in this process of cultural transmission have been played down in the process of asserting the predominance of an exclusively 'western tradition', with all its geographical and intellectual connotations of Christian, European authority and superiority. However, in many areas scholars had to deal with classical texts which had survived as a result of their cultivation by Arab scholars. From the ninth to the fourteenth centuries the most innovative scholarly activity within the fields of astronomy and cosmography emanated from Arabic centres of learning in North Africa, Moorish Spain and the territories increasingly under the control of the Ottomans. Many of the early Portuguese forays into the fields of astronomy and navigation emerged directly from the complex inheritance of Islamic and Hebraic Iberian scholars.[16] The manuscript texts of Ptolemy's astronomical and geographical works were no exception to this process of intellectual transmission and circulation. Throughout the ninth and tenth centuries most of the classical works of science and mathematics had been translated into Arabic at the Academy of Science established in Baghdad in 813 by al-Ma'mun, the Caliph of Baghdad.[17] The academy was particularly active in the field of astronomy as well as the attempt to calculate the precise length of a degree in the wake of Ptolemy's studies. Arab scholars such as al-Ma'mun not only preserved classical scholarly texts, but in many instances built on such

98

knowledges to develop new scientific theories. Texts included not only Ptolemy's *Geographia* but also his astronomical treatise the *Almagest*, which was translated into Latin from Arabic for the first time in 1175 by Gerard of Cremona, whose translation formed the basis of the first printed edition of the text in Europe in 1515.[18] As for the *Geographia*, it was subsequently transcribed into Syriac, and formed the basis for most of the geographical and cosmographical research which proceeded to emerge from al-Ma'mun's academy.[19]

Nor was the emergence of Ptolemaic texts limited to their circulation within Baghdad. There remain thirteen known works produced in Arabic in manuscript prior to the twelfth century, including three copies of the *Geographia*.[20] However, historians of cartography still insist on eliding this aspect of the transmission of Ptolemy, in favour of stressing a neat linear movement of the text from Byzantium to Italy. Such a view disregards the extent to which the territorial growth of the Ottoman empire throughout the fifteenth century gradually assimilated ancient centres of Arabic learning into its imperial structure. In 1471 the renowned astronomer ᶜAlī Qushjī, the head of the observatory in Samarkand, moved to Istanbul to continue his studies under the generous financial benefaction of Mehmed. The substantial library which he took with him (and which undoubtedly included copies of Ptolemy's work) transformed the field of Ottoman cosmography.[21] With ᶜAlī Qushjī came a whole host of scholars as well as artists and architects, drawn to Istanbul as Mehmed sought to redefine the city as an Islamic Constantinople, a cultural capital fit to rival the great civic centres of fifteenth-century Italy. Throughout the 1470s and 1480s Italian artists and craftsmen of the stature of Gentile Bellini, Constanzo de Ferrara, Matteo de' Pasti and the architect Filarete were all reported to have played an active part in culturally restructuring the Ottoman capital.[22] Istanbul became a major centre for intellectual and artistic production, which was enthusiastically supported by a sultan keen to emphasize that his cosmopolitan capital was at the centre, rather than on the periphery, of a whole range of artistic and scholarly traditions and initiatives. Nowhere was this situation more vividly illustrated than in Mehmed's patronage of geographical research. Scholars such as ᶜAlī Qushjī provided a vigorous intellectual context for the study of cosmography and geography, and one particularly striking instance of Mehmed's subsequent interest in this area underlines Berlinghieri's desire to dedicate his *Geographia* to the Ottoman sultan.

One of the most respected scholars amongst Mehmed's retinue was the Greek chronicler Kritovoulos of Imbros. In his manuscript *The History of Mehmed the Conqueror*, written in honour of the Ottoman sultan's achievements, Kritovoulos claimed that:

His highness the Sultan used to read philosophical works translated into the Arabic from Persian and Greek, and discuss the subjects of which they treated with the scholars of his court. Having read the works of the renowned geographer Ptolemy and perused the diagrams which explained these studies scientifically, the Sultan found these maps to be in disarray and difficult to construe. Therefore he charged the philosopher Amirutzes with the task of drawing a new clearer and more comprehensible map. Amirutzes accepted with pleasure, and worked with meticulous care. After spending the summer months in study and research, he arranged the sections in scientific order. He marked the rivers, islands, mountains, cities and other features. He laid down rules for distance and scale, and having completed his studies he presented the Sultan and those engaged in scholarship and science with a work of great benefit. Amirutzes wrote the names of the regions and cities in the Arabic script, and for this purpose engaged the help of his son, who was master of both Arabic and Greek.[23]

Georgius Amirutzes worked on his new version of Ptolemy throughout the summer of 1465, and appears to have been helped in his new commission by George Trepuzuntios, another scholarly admirer of Mehmed. Trepuzuntios had returned to Istanbul from Italy in the same year to discuss with Amirutzes his new Greek translation of Ptolemy's astronomical treatise the *Almagest*, a text which Trepuzuntios proceeded to dedicate to Mehmed. In his *History* Kritovoulos went on to note that Mehmed subsequently 'ordered him [Amirutzes] to issue the entire book in Arabic, and promised him large pay and gifts for this work'.[24] This remarkable commission was completed by Amirutzes towards the end of 1465, and a surviving manuscript version of the map offers testimony to the skill and learning which clearly went into its production (illus. 32). Despite the initial unfamiliarity of its appearance, with its Arabic nomenclature and south oriented to the top of the map, its detailed representation of the Ptolemaic *oikoumene*, along with its sophisticated conical projection complete with scales of latitude, emphasizes that it was one of the most up-to-date fifteenth-century representations of the globe based on Ptolemy's calculations. It should therefore be stressed that the apparently alien appearance of the map can be ascribed to the systematic occlusion of such traditions of Islamic mapping within histories of cartography, and not to its novelty as an isolated example of Ottoman-sponsored geography simply aping the conventions of Italian mapmaking.

Mehmed's sponsorship of Amirutzes' redrawing of Ptolemy's *oikoumene* establishes the fact that Ottoman scholars were able to draw on a body of Arabic scholarship when working on a text such as Ptolemy's *Geographia*, a text to which they had as much intellectual claim as the scholars of fifteenth-century Italy. The recovery of shared classical texts was pursued as vigorously in Istanbul as it was in Italian centres such as Florence and Urbino. In fact, Kritovoulos' comments suggest that the Ottoman

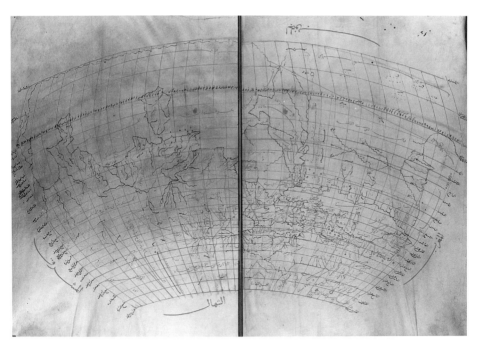

32 Georgius Amirutzes, world map, *c.* 1465, vellum. Ayasofya Library, Istanbul.

scholars, and even Mehmed himself, had a more comprehensive range of both Greek and Arabic copies of Ptolemy to consult in the construction of their new editions, and that this range of intellectual reference allowed Ottoman geographical scholarship to develop along lines not dissimilar to those developed in Italy around the same time. Figures such as Borso d'Este, the Duke of Ferrara, whose family was responsible for secretly obtaining the Cantino Planisphere from Lisbon in 1502, financially supported the work of geographers like the German scholar Donnus Nicolaus Germanus. Donnus Nicolaus' 1466 Latin manuscript of Ptolemy's *Geographia* was subsequently dedicated to the duke and he received 100 gold ducats for it.[25] The pattern of patronage established for the production of new geographical texts, and adapted throughout the courts of fifteenth-century Italy, operated on very similar terms to those laid down by Mehmed in Istanbul in the late fifteenth century, a fact which has on the whole been disregarded in critical accounts of the place of the Ottomans in the culture of early modern Europe.[26]

Both the Amirutzes commission itself and the subsequent model of patronage employed by Mehmed can be seen to breach the perceived geographical and intellectual boundaries of the European Renaissance, which have excluded Ottoman culture, even more than Arabic culture, as exhibiting little concern for scholarly learning and thus completely divorced them from the formation of 'western' cultural values. Mehmed's patronage of Amirutzes' Ptolemy exemplifies the knowing participation of the Ottoman court in the transmission and ultimate shape of classical texts subsequently judged to be crucial in the formation of the European western tradition. This underlines the extent to which the European inheritance of the Graeco-Roman tradition was equally suffused with the achievements of Arabic scholars based firmly within the bounds of the Ottoman empire. It can only be concluded that with the rapid exchange of scholars and artists between Istanbul and Italy towards the end of the fifteenth century, Berlinghieri was only too aware that, both in terms of political power and intellectual inheritance, Mehmed was the most appropriate patron for his *Geographia*.

Subsequent clandestine exchanges of geographical material between Italy and Istanbul suggest that scholars and diplomats other than Berlinghieri were aware of Mehmed's scholarly interest in geography, and that the Ottoman sultan was also cognizant of the military and diplomatic importance of maps to the expansion of his already vast empire. Matteo de' Pasti's visit to Istanbul in 1461 at the behest of Sigismondo Malatesta, mentioned earlier, offers a more covert account of the traffic in geographical information. De' Pasti's journey was ostensibly a cultural visit designed to assist Mehmed in the building of his new imperial palace. Off Crete, de' Pasti was captured by Venetian officials, who returned him to Venice where he was accused of being a spy in league with the Ottomans. In his possession the authorities found Sigismondo's letter to Mehmed, in which the Lord of Rimini announced:

The greatness of your majesty is such that its formidable power easily exceeds the capacity not simply of myself, but of the whole human race. It seemed to me, therefore, that if what I should offer was to be of sufficient stature it should be something close to me personally, and unique to me, something which I believed would thereby give you the greatest pleasure. So I have decided to make you companion to my studies and my pleasures, and to share with you a substantial and erudite work on military matters, in which are to be found the great leaders and imperial rulers not simply of our own time, but of preceding ages, all of whom, incontrovertibly, and with one voice you are acknowledged to excel and surpass in merit.[27]

The manuscript which Sigismondo delegated de' Pasti to present to Mehmed was Roberto Valturi's highly influential *De re militari*, an illustrated work on military arms and tactics. Accompanying it was another document of crucial diplomatic importance – a detailed manuscript map of Italy.[28] Contemporary accounts suggest that this map contained logistical and military information vital to any subsequent Ottoman advance on Venice and the Italian peninsula. Sigismondo's anti-Venetian feelings were well known, and it appears that he was prepared to encourage Mehmed to invade Italy with the offer of detailed cartographic information which would be of great value to the authorities in Istanbul. Such threats were not taken lightly by the Venetians, and their response to de' Pasti's breach of security was validated in the light of Mehmed's eventual invasion of Italy in the summer of 1480, when Gedik Ahmed Pasha landed a force of 18,000 soldiers at Otranto. A badly mutilated map of the Venetian territories specifying vulnerable ports and military fortifications, dated to the 1470s and still in the possession of the imperial library in Istanbul, suggests that other clandestine maps managed to find their way successfully into the hands of the Ottoman court as it planned its assault on Italy in the final years of the 1470s.[29]

The traffic in maps, both overt and covert, and geographical information between Italy and Istanbul is symptomatic of the centrality of the Ottoman empire to the social and political dynamics of the evolving world of early modern Europe. Its court appears to have been well aware of the significance of appropriating and utilizing classical learning in the service of political action, and played as crucial a role in the recovery of classical texts like Ptolemy's *Geographia* as the city-states of fifteenth-century Italy. The elaborate commissioning, and often clandestine appropriation, of a range of geographical material by the Ottoman court also suggests that a shared perception of the status of cultural production and the importance of the mechanics of diplomatic exchange existed between Istanbul and the city-states of Italy. This perception encompassed cultural exchanges predicated on amity, as in the case of Berlinghieri's attempt to court Mehmed, as well as exchanges predicated on potential acts of aggression, as in the case of Malatesta's attempt to smuggle a map of Italy into the Ottoman court. Neither exchange was predicated on the perception of the Ottomans as mysterious, demonic others, anterior to the culture of a confidently self-defined European polity. Such exchanges were predicated on the belief that the Ottoman empire was a powerful neighbour with whom it was possible to do both diplomatic and commercial business, and whose political concerns often coincided with those of the city-states of fifteenth-century Italy.

Sea Matters: Ottoman Cartography in the Mediterranean

The maps discussed so far invariably concerned the mapping of land and strategic locations across which diplomats and armies moved like pieces on an enormous geographical chessboard. The difficulties associated with keeping track of the movements of ever larger armies in time and space, and the problems which emerged from the administration of the increasingly large tracts of territory that began to come under the control of centralized imperial authority, made the need for such territorial maps imperative. However, they remained something of a novelty even by the end of the fifteenth century. To some extent this explains the relative scarcity of the maps already discussed. The fact that the Ottoman authorities were quick to capitalize on the scarce geographical resources that were available, in order to facilitate the administration and potential domination of the enormous territories to which they laid claim, only emphasizes their acuity in swiftly perceiving the political importance of maps and geographical information. Nevertheless, the commercial rhythms which defined the maritime world of the Mediterranean continued to operate with little regard for the diplomatic machinations which took place on land. The manuscript portolan charts which enabled pilots and merchants to navigate throughout the waters of the Mediterranean changed very little throughout this period. This was partly as a result of the realization that the logistical capabilities of such charts were perfectly matched to the commercial requirements of the sailors and traders who regularly moved between the islands and ports which made up the Mediterranean world. There was also very little need for innovation as, unlike contemporaneous Portuguese navigators who moved along the coast of Africa, Mediterranean pilots rarely had to navigate across uncharted territory which potentially questioned or compromised the effectiveness of the standard portolan charts.

Historians of cartography have endlessly debated the origin of these charts, which date back to at least the twelfth century. Recent critical consensus has argued that the portolan tradition probably emerged from either Genoa or Catalonia, two of the most prominent trading centres within the Mediterranean.[30] However, as with cartographic histories of the transmission and circulation of Ptolemy's texts amongst the scholarly élites of the early modern world, this assumption has invariably neglected the importance of other early centres of scholarly learning and portolan chart production. A significant amount of early charts were produced in the North African cities of Tunis and Tripoli.[31] One of the earliest extant charts produced in Arabic, known as the 'Maghreb chart' (illus. 33), has been dated 1330 and is believed to have been produced in Granada or Morocco. Drawn by an anonymous Arabic mapmaker, it contains all the standard traits of the early portolan chart, with its proliferation of rhumb lines, exact delineation of coastal territory and place names running perpendicular to the coastline. However, the toponymy tells an interesting story as to its cultural origin. Of the 202 place names that are identifiable, 48 are of definably Arab origin, whilst the rest are made up of Catalan, Hispanic and Italian.[32] The fact that even charts made throughout the ports of

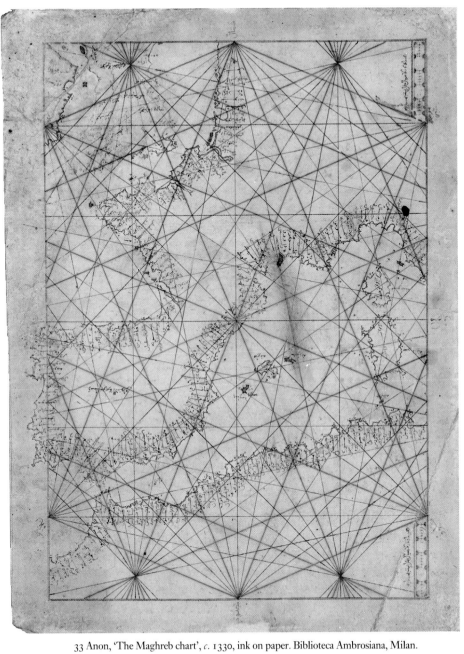

33 Anon, 'The Maghreb chart', *c.* 1330, ink on paper. Biblioteca Ambrosiana, Milan.

Italy also consist of places and terms derived from Arabic emphasizes their hybrid nature, suffused with as much Arabic influence as Catalan or Italian. Nor is this surprising considering the amount of influence which Arab communities at both ends of the Mediterranean exerted on the commercial and cultural life of the area. The rise to political pre-eminence of the Ottomans within the Mediterranean by the fifteenth century also inevitably left its mark on the portolan charts of the period. Amongst its extensive collection of fifteenth-century charts and sailing atlases, the Topkapi Saray Museum in Istanbul still retains a magnificent chart of the entire Mediterranean made by Ibrāhīm ibn Ahmad al-Kātibī in 1413–14 (illus. 34). Prominently displaying the castle of Tunis, where it was produced, and comprehensively labelled in *Maghribi* script, the chart appears to have come into the possession of the Ottoman imperial authorities by the beginning of the sixteenth century, an example of the type required by both Italian and Arabic pilots as they sailed throughout the sea lanes of the Mediterranean. Produced by Arabic geographers and pilots, such charts shared the geographical coordinates employed by Catalan and Genoese ones, which also presumably drew on the knowledges contained in the Arabic charts as they circulated throughout the Mediterranean.

The predominantly commercial exchanges which gave rise to these charts meant that the navigational *lingua franca* of the Mediterranean included Arabic and, increasingly, Turkish components, underlined by the fact that Italian and Catalan merchants did the majority of their business in commercial centres under Islamic jurisdiction, such as Tripoli, Tunis, Alexandria and, increasingly, Istanbul. Even charts themselves were traded as valuable merchandise by merchants who held factors throughout the Mediterranean. Between 1390 and 1392 Domenach Pujol, a merchant based in Barcelona, entrusted a collection of portolan charts to one Pere Jalbert, a mariner and factor in his pay, who was given instructions to sell them throughout the ports of the Mediterranean, including Genoa, Naples, Sicily and Alexandria.[33] In Alexandria Jalbert was instructed to barter the charts for consignments of pepper, a striking example of the ways in which these charts moved between cultures as valuable commodities in their own right, exchangeable objects whose value could match the price of a cargo of pepper. The hybrid *lingua franca* of the region suggests that it is unlikely that they were only of interest to Christian pilots, but that they were also of use to the Arabic pilots who similarly navigated their way across the waters of the Mediterranean. The quest for the cultural origins of the portolan chart should therefore be regarded as a futile enterprise, precisely because the Mediterranean world was a shared commercial and intellectual space whose different cultural influences combined to produce trading mechanisms and objects, such as these charts, which were not

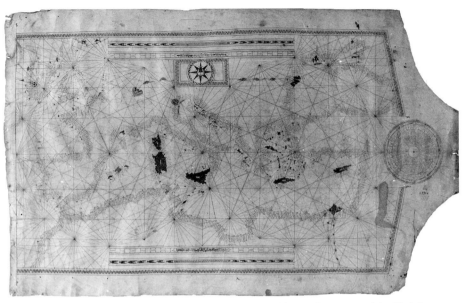

34 Ibrāhīm ibn Ahmad al-Kātabī, portolan chart of the Mediterranean and the Black Sea, 1413–14, vellum. Topkapi Saray Museum.

unique to any one culture or area but emerged as a result of the diverse range of transactions which took place between these cultures. To deny the presence of Arabic and, increasingly, Turkish involvement in the evolution of the portolan chart only reproduces the ideologically weighted critical belief that the Ottomans played no part in the intellectual development of more scholarly geographical texts like Ptolemy's *Geographia*.

The spectacular rise of the Ottoman empire to political and commercial pre-eminence in the eastern Mediterranean towards the end of the fifteenth century ensured that the balance of power in the region altered significantly. The territorial conquests made under Mehmed II were consolidated during the reign of his son Bayezid, who also ordered the building of an enormous navy to ensure Ottoman control over the coastal areas of its empire. By the winter of 1500 the navy was said to number around 400 vessels, which included 200 galleys fitted with heavy cannon, located at bases in Galata on the Black Sea, Avlonya on the Adriatic, Chios in the Aegean and Macri in the eastern Mediterranean.[34] Unmatched throughout the Mediterranean in terms of numbers and firepower, the fleet's responsibilities focused primarily on the protection of shipping lanes and the establishment of Ottoman-dominated seaborne trade routes, rather than outright maritime aggression. Throughout the early years of the sixteenth century it provided the ailing Mamluk regime in Egypt with protection from Portuguese incursions into Mamluk territories. However, this strategic use of the fleet was based on a more wide-ranging plan to extend Ottoman influence into the Mamluk territories and ultimately control the commercially vital shipping lanes of the Red Sea, which brought the majority of merchandise from Asia into mainland Europe. By 1517 the Ottomans had annexed Egypt, as well as Syria and Yemen, as part of a wider attempt to gain easier access to the seaborne trade routes to India which were being increasingly monopolized by the Portuguese. Not surprisingly, these dramatic advances in Ottoman naval technology and power were supported by a whole array of geographical and navigational information required by the fleet in the attempt to establish its dominance over both the eastern Mediterranean and the Red Sea. One of the commanders of the fleet used in the annexation of Egypt in 1517 was also responsible for some of the finest maps and charts of the Mediterranean produced in the early years of the sixteenth century. Simply known as Piri Reis, in 1517 he presented a manuscript map of the world (illus. 35) which he had completed in 1513, to Sultan Selim I in Cairo. In his account of this presentation the Turkish commander recalled that:

I, your humble pilot, made maps in which I was able to show twice the number of things contained in the maps of our day. Having made use of new maps of the Chinese and the Indian Seas which no one in the Ottoman lands had hitherto seen or known, I presented them to the late and deceased Sultan Selim while he was in Egypt and received his favour.[35]

Whilst only a fragment of the western portion of this map still survives, the notes which Piri Reis appended to it emphasize the enormous variety of geographical knowledges to which he had access:

No such map exists in our age. Your humble servant is its author and brought it into being. It is based mainly on twenty charts and *mappa mundi*, one of which is drawn in the time of Alexander the Great, and is known to the Arabs as Caferiye [*dja'grafiye*]. This map is the result of comparison with eight such [*dja'grafiye*] maps, one Arab map of India, four new Portuguese maps drawn according to the geometrical methods of India and China, and also the map of the western lands drawn by Columbus; such that this map of the seven seas is as accurate and reliable as the latter map of this region.[36]

There appears to be little doubt that Piri Reis consulted Arabic, Greek and Italian copies of Ptolemy's *Geographia* available to him in Istanbul, as well as more recent portolan charts of both Portuguese and Indian provenance. As a naval commander with exposure to the scholarly geography of Ptolemy, as well as practical involvement with voyages which brought him into contact with both Portuguese portolan charts and the local charts of the Indian Ocean which were so coveted by the Portuguese

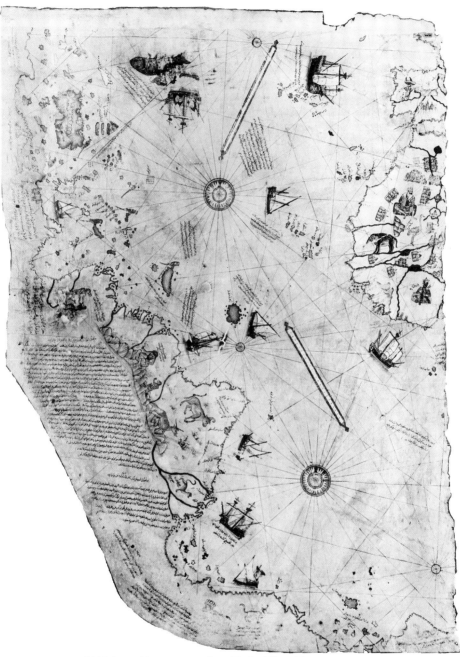

35 Piri Reis, world map, 1513, vellum. Topkapi Saray Museum.

crown, it would be no exaggeration to claim that by 1517 Piri Reis had access to one of the most comprehensive collections of geographical information available to any scholar or pilot active in the early modern world. His world map was not only one of the most detailed to be produced in the first decades of the sixteenth century. It was also one of the first maps to account systematically for the significance of Columbus' voyages to the Americas. That Piri Reis should present such a map of the world to an Ottoman sultan, himself intent on imperial expansion on a global scale, clearly refutes any notions that the Ottomans took no interest in geographical matters or the global expansion of other empires within the early modern world.

Nor did Piri Reis' cartographic activities end there. In the ensuing years the Turkish commander set to work on compiling one of the most comprehensive collections of portolan charts of the Mediterranean to be produced in the early sixteenth century. By 1521 Piri Reis had completed his manuscript. Entitled *Kitāb-i baḥriye* (translated as 'Manual of maritime or sea matters'), it was a comprehensive account of the sailing directions required to navigate throughout the Mediterranean. By 1526 Piri Reis completed an even longer version of the text. It contained 216 exquisitely drawn portolan charts of the region (illus. 9) accompanied by 130 chapters of detailed hydrographic description of islands, coastlines, and fortifications which Piri Reis had traversed throughout his career with the Turkish fleet. In the same year he presented the completed manuscript to Sultan Süleyman the Magnificent. However, the Turkish admiral did not draw the geographical limit of his cartographic enquiries at the Mediterranean alone. In a radical move indicative of the extent to which the portolan chart began to adapt to the expanding political and commercial parameters of the early modern world, Piri Reis presented Süleyman with a copy of the *Kitāb-i baḥriye* complete with significant new additions to the first manuscript, which were to be of particular interest to the new sultan. These offered a detailed outline of the development of Portuguese exploration from the first voyages down the west coast of Africa in the fifteenth century right up to the contemporary establishment of Portuguese commercial hegemony in the Indian Ocean. With remarkable accuracy Piri Reis recounted the exploits of Bartolomeu Dias and the subsequent discovery and geographical position of the Cape of Good Hope:

He was the one who found the land of the Abyssinians and the Cape of Good Hope.
For what they call *Kavu Bono Ispiranse* is what we refer to as *Umiz Burnu*.
They sought this place mile by mile and discovered it in the ninth year.
This place is the beginning of the route to India and this is why they made it their goal to find it.

Every year they advanced a thousand miles. At the farthest place they reached, they set up a marker and returned.

When the sailing season returned, they would come back and seek this marker. And from there, they immediately set out to go further but they were unable to find the open sea.

So as not to be misled on their course, they also traversed the shores on foot.

In this way they measured the land, furlong by furlong, advancing on their way.

Those who discovered *Kavu Bono Ispiransa* set up there a column and a crucifix.

They held a great celebration there, for they had found the way to India.

Know then that this *Umiz Burnu* is the farthest extremity of the continents of the world.

This place is part of the southern region and it lies beyond the Equator.

That cape lies thirty-four degrees below the Equator.[37]

The remarkable accuracy of Piri Reis' geographical information was matched only by his extensive knowledge of the development of maritime Portuguese cartography throughout the fifteenth and early sixteenth centuries. This was reflected in his highly accurate assessment of the relationship between cartography and the successful commercial expansion of the Portuguese crown:

In Portugal there was an influential priest who was thoroughly versed in all things.

Skilled in theoretical and practical knowledge, he had attained perfection.

Through constant effort he had become a philosopher.

He created a ball shaped like an apple on which he marked the lands and the seas.

On that globe he indicated all the distances and calculated its size to be twenty-four thousand miles.

He marked the various countries and wrote their names on the globe.

And at the same time he recorded on it the islands in the seas and countless numbers of castles.

Using this globe, one can travel the whole world across the Ocean, as far as China.[38]

The 'influential priest' appears to be a reference to Fra Mauro, whose world map commissioned by the Portuguese crown reached Lisbon in 1459. The 'ball shaped like an apple' refers to Martin Behaim's famous terrestrial globe, used by Magellan in his circumnavigation of the globe in 1522, and argued to have influenced Columbus' voyage of 1492. The detail of Piri Reis' account suggests an acquired knowledge of not only the geographical artefacts associated with Portugal's seaborne expansion, but also an awareness of early Portuguese chronicles which narrated the history of this expansion in terms strikingly similar to the description offered by Piri Reis. However, to read his comments as the awed response of someone unfamiliar with such cartographic developments does not accord with the evidence of Ottoman navigational advances made during this period. Such advances suggest that Piri Reis may have been rather puzzled by the limitations of Portuguese cartography. The accuracy with which Arabic pilots navigated throughout the Indian Ocean, using a variety of instruments from which they measured the altitude of the sun and the position of the stars (techniques clearly known to Piri Reis from his own naval experiences throughout both the Mediterranean and the Indian Ocean), must have made the much-vaunted maps and globes of Fra Mauro and Behaim seem disappointingly simplistic to someone as geographically experienced as the Turkish commander.

As a lavishly designed text ceremonially presented to Süleyman the Magnificent, Piri Reis' *Kitāb-i baḥriye* appears to have followed in the path of earlier examples of Ottoman cartography as both an aid to navigational expertise and a prized political object, part of the diplomatic apparatus utilized by the Ottoman court in the pursuit of an ambitiously expansionist foreign policy. Piri Reis was, like his contemporaries in the Italian mapmaking industry, only too aware of this dual function which his charts and geographical observations served, as his comments on their value to those in the diplomatic profession suggest:

... those who are masters of this profession may by applying that which is written in this book and with the grace of God facilitate all their affairs, even if they have never seen or been acquainted with such places, and they will have no need of pilots.[39]

Whilst the claim that the Ottoman authorities would 'have no need of pilots' in their utilization of the *Kitāb-i baḥriye* may have been rather optimistic, Piri Reis' comments suggest that he was convinced of the ability of his charts to facilitate diplomatic, commercial, naval and military logistics to the advantage of the empire. With remarkable insight he realized that the intellectual and commercial parameters of the Mediterranean world were becoming increasingly diffuse with the impact of both Columbus' voyages to the Americas and da Gama's expedition to India, and that any collection of portolan charts based on the Mediterranean had to reflect this level of geographical diversity. As a result, his text was not only a significant Ottoman addition to the culturally hybrid tradition of Mediterranean portolan chartmaking, but it was also partly responsible for resituating the portolan chart within a wider commercial and geographical context than had previously been the case.

The central cause of this widening of geographical focus emanated from the changing commercial dynamics of the Mediterranean world which proceeded Portugal's successful establishment of the trade route to India via the Cape of Good Hope. Piri Reis' *Kitāb-i baḥriye* reflected this shift in commercial and political awareness, and emphasized the extent to which the Ottomans were aware of the need to produce geographical and hydrographical information which kept up with the pace of change in the Mediterranean world and beyond. The manuscript was therefore also a valuable source of information on the effects which the Portuguese were having on the trade throughout the Indian Ocean, which, by extension, affected the nature of trade in the Mediterranean. In his description of Hormuz, whose strategic location at the entrance to the Persian Gulf made it a vital conduit in the movement of merchandise between both the Indian Ocean and the Mediterranean, Piri Reis noted:

But now the Portuguese have reached this place and they have built a fort on that Cape.
There they wait and collect tolls from ships that pass. You have now learned the circumstances of this place.
The Portuguese have overcome them all and the khans of the island are filled with their merchants.
In fall or summer, no trading takes place unless the Portuguese are present.[40]

The gathering of geographical information for the completion of the *Kitāb-i baḥriye* can be seen here as coterminous with the collection of commercial information on the adverse effect that strategic Portuguese possession of commercially vital cities and ports such as Hormuz was having, not only on the revenues of the Ottoman empire but also on the trade of the Mediterranean more generally. The Portuguese capture of Hormuz in 1515 did signal a measure of Portuguese pre-eminence in the trade throughout the Indian Ocean, in practice forcing local traders to purchase *cartazes* which obliged them to pay customs duties on all merchandise direct to the Portuguese authorities.

It is bitterly ironic that it was Piri Reis' ability to keep the Ottoman court abreast of these developments in the Persian Gulf that was to lead ultimately to his death. In 1547 he was appointed admiral of the Ottoman fleet by Süleyman, with orders to oppose the growing commercial and naval power of the Portuguese in the Indian Ocean. In 1549 he completed the first stage of his mission by taking the city of Aden. By 1552 the Turkish admiral was laying siege to Hormuz, but with the arrival of Portuguese naval reinforcements he was forced back to Egypt. Forbidden to return to the Persian Gulf for his failure to take Hormuz, he was beheaded in Cairo in 1554 on the orders of his former geographical patron, Süleyman the Magnificent.[41] Whilst Piri Reis' cartographic achievements arguably made Ottoman naval confrontation with the Portuguese in the Persian Gulf possible, it would seem that his geographical acumen could not save him in the face of naval defeat. Nevertheless, his *Kitāb-i baḥriye* made an enormous, if historically neglected, contribution to the tradition of Mediterranean portolan chartmaking. The information gathering which went into these charts suggests that a much more fluid exchange of hydrographical and commercial information brought such charts into being than has previously been conceded by historians of cartography. Piri Reis' text is littered with comments signifying that his information was obtained athwart the political divisions which ostensibly structured the diplomatic confrontations of the period. In his account of the Portuguese involvement in China and the trade in porcelain, he notes that 'I asked these Portuguese about this land of China'.[42] As has been seen, at the same time the Portuguese were busy gathering nautical information on the sea routes between East Africa and China from Muslim and Javan pilots. A level of geographical exchange was apparently taking place which circulated information on navigation, topography and commerce that far exceeded the traditional political distinctions which historians have claimed informed the public rhetoric of both Ottoman and Christian authorities as they struggled for commercial and political pre-eminence in the Mediterranean and the Indian Ocean. Existing between the practical portolan charts used throughout the early sixteenth-century Mediterranean and the diplomatically valuable copies of Ptolemy circulated between the élites of Florence, Venice and Istanbul, Piri Reis' *Kitāb-i baḥriye* signified, in both its political status at the Ottoman court and its geographical range of reference, that the perception of the Ottomans and their 'eastern' empire as the exoticized, mysterious and feared antithesis to the classical civic humanism of Renaissance Italy was a myth. The cultural and commercial transactions mediated through scholarly geographical texts such as Ptolemy's *Geographia* and Piri Reis' portolan charts expose the fallaciousness of arguments claiming that the Ottomans were geographically isolated from the political and diplomatic contexts of 'Renaissance Europe'. Such transactions also give the lie to the traditional belief that the Ottomans were indifferent to the cultural production and commercial exchanges which took place throughout the Mediterranean.[43]

Other Geographies?

Whilst Piri Reis' geographical activities resulted in perhaps the most spectacular and comprehensive examples of early sixteenth-century Ottoman cartography, they were not the only ones produced at this time. The continuing political influence of the Ottoman empire on sea and land throughout the sixteenth century stimulated a whole range of geographical initiatives developed by Turkish scholars, architects, artists and diplomats.[44] Having absorbed the advantages of utilizing territorial maps such as those contained within Ptolemy's *Geographia* in envisioning territorial conquest, Süleyman the Magnificent engaged the services of the

Turkish topographer Maṭrāḳčī Naṣūḥu to record the progress of his military campaign against the Safavids between 1533 and 1536. The subsequent manuscript atlas of the campaign, completed in 1537 and entitled *Beyāni-i Menāzil-i Sefer-i Irākeyn-i Sultan Süleyman Ḫān*, contained 107 elevated landscape miniatures of Süleyman's campaign. So successful was the atlas in establishing a visual rhetoric of territorial conquest that Maṭrāḳčī was commissioned again in 1542 to record Süleyman's territorial and naval attacks on his imperial adversary the Habsburg emperor Charles V. Entitled *Tārih-i Fetḥ-i Şiḳlōş ve Estergon ve Istūnibelġrād*, Matrakçi's topographical plans contained depictions of Süleyman's march on Hungary, as well as miniature plans of Mediterranean ports which came under Ottoman attack throughout 1542 and the subsequent year. These included maps and plans of Toulon, Marseilles, Antibes, Genoa and Nice[45] (illus. 36).

Nor did Süleyman's campaigns on land result in the waning of Ottoman nautical cartography. By the late 1550s, prior to Süleyman's death in 1566, a collection of sea atlases which included comprehensive maps of the world had been produced by mapmakers based in Istanbul. The anonymous *Deniz atlasi*, tentatively dated to 1560, not only contains a series of portolan charts of the Mediterranean and the Black Sea, but also a detailed chart of the Indian Ocean, as well as a map of the world (illus. 37). Its successor, the so-called 'Ali Mācār Reis atlas, dated 1567, contains six portolan charts and a world map with similarities to the *Deniz atlasi*. As Mācār means Hungarian in Turkish, it seems likely that its author, 'Ali Mācār, was in fact a Hungarian geographer working for the Ottoman authorities, whose charts drew on his experience of both Ottoman and European regions to create a comprehensive atlas focused primarily on the Mediterranean. By 1570, the same year that Abraham Ortelius published his monumental *Theatrum Orbis Terrarum*, another anonymous Turkish geographer completed the so-called *Atlas-i hümayun*, which appears to have drawn on both the *Deniz atlasi* and the 'Ali Mācār Reis atlas.[46]

But perhaps no better example of the scale and extent of Ottoman cosmographical and geographical investigation can be invoked than the Ottoman miniature depicting the activities of the Turkish court astronomer Takiyüddin, at work in his observatory in Istanbul in the 1580s (illus. 8). The observatory was built in 1575 with the support of Takiyüddin's imperial patron, Murad III, and acted as the focus for the geographical and celestial enquiries of the astronomer and his fellow cosmographers.[47] Lining the back wall are precious manuscripts (including, no doubt, copies of Ptolemy) of the type being consulted by Takiyüddin and his assistants. In the middle ground cosmographers take measurements with astrolabes, quadrants and compasses, whilst in the foreground stands a terrestrial

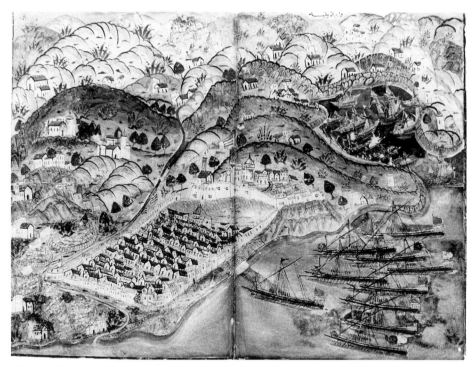

36 Maṭrākçī Naṣūḥu , Nice, from the *Tārih-i Fetḥ-i Şiķlöş ve Estergon ve Istūnibelġrād, c.* 1543.
Topkapi Saray Museum.

globe, with Europe and Asia clearly marked, which appears to be based on
the geographical information contained in the Turkish atlases of the previous decade. This illustration of Takiyüddin's activities is symptomatic of
the intense level of geographical enquiry that took place in Istanbul from
its establishment as the Ottoman capital in 1453. Such activity has, however, been marginalized in accounts of early modern geography due to its
connection to an empire which was demonized as simply concerned with
the acquisition of territory in the pursuit of a global Islamic imperialism.
Such accounts fail to recognize the sheer diversity of Turkish geographical
thinking, and the extent to which such thinking interfaced with similar enquiries which were taking place throughout mainland Europe at exactly
the same time. Perhaps the fact that, like the illustration of Takiyüddin's
observatory, all the examples of Turkish cartography which have been analysed were produced in manuscript form, provides one final indication as
to their continued occlusion within critical accounts of the Renaissance.
The impact of printing on the culture of early modern Europe was without doubt one of the most decisive factors in shaping the political and
imaginative parameters of the early modern world. However, the focus
upon print as the driving force behind most forms of intellectual develop-

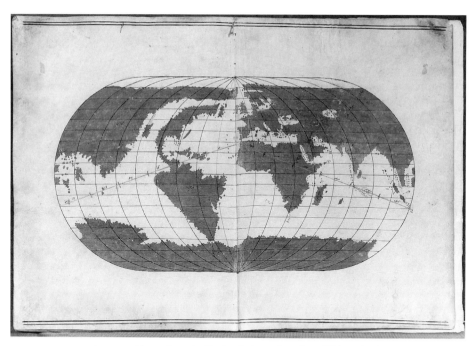

37 Anon, world map from the *Deniz atlasi, c*, 1560, vellum. Walters Art Gallery, Baltimore.

ment has been at the expense of cultures and traditions which continued to express their intellectual output through the medium of the manuscript. As the culture of print was not assimilated into the Ottoman empire until the beginning of the eighteenth century, it is easy to see why critics have been much more reluctant to assess the importance of Ottoman learning within the specificities of scribal production. The fact that one of the first texts to roll off the printing press of the founder of Turkish printing in 1726 was a map of the Ottoman empire is a poignant reminder of the prestige accorded to the study of geography within the empire.[48]

In tracing the circulation of Turkish maps and charts and their interface with both the Ottoman court and the territories over which it held sway, it becomes possible to define a level of territorial and cultural interaction between Istanbul and its western European counterparts which breaches the artificial geopolitical boundaries of 'East' and 'West'. The sovereigns and city-states of early modern Europe were neither politically powerful nor diplomatically cohesive enough to establish a sustained and coherent demonization of the Ottoman empire. At various times the Ottomans stood as implacable foes, at others as potential allies. In tracing the transactions of knowledge which produced the remarkably diverse collections of maps and charts which characterized political relations throughout early modern Europe and the Ottoman empire, it is possible to offer a very different account of the relations between communities than is reflected in the contemporary statements of princes and diplomats, which invariably painted a picture of ideological conflict and religious antagonism. As a result, the study of early modern Europe is perhaps now at the stage where there is a possibility, as Edward Said has recently argued, 'of re-thinking and reformulating historical experiences which had once been based on the geographical separation of peoples and cultures'.[49] This is certainly the case in critical perceptions of the relations between early sixteenth-century Europe and the Ottoman empire.

In redrawing the map of the early modern world in this way, the status of maps themselves can be similarly reconsidered by following their various diplomatic and commercial trajectories as they moved between locales such as Florence, Venice, Istanbul and Cairo. The portolan maps and charts of Piri Reis, and the later Turkish atlases, graphically portray the complex relations between communities throughout the Mediterranean, and reflect how it becomes impossible to delimit them into simple categories of Christian or Muslim, European or Turk. The geographical texts of the likes of Berlinghieri and Amirutzes also point to the status of scholarly maps as skilfully marketed luxury objects which could be offered to, or produced within, the powerful Ottoman court. Imbued with arcane diplomatic and commercial knowledges, which only served to enhance their status as valued objects, such maps were politically and aesthetically appreciated by Christian and Muslim owners alike. As a result, it becomes impossible to sustain the Orientalist myth of the exotic, peripheral figure of the despotic Turk, which has implicitly informed so many accounts of the development of early modern Europe.[50]

4 Cunning Cosmographers: Mapping the Moluccas

The previous two chapters circulated around the diplomatic and commercial disputations which saw the great empires of the early modern world struggling for nautical and commercial supremacy over the prized trade routes which crisscrossed the Indian Ocean. By 1517 the Portuguese had established a monopoly on the spices, pepper and other precious goods as far east as Malacca by the strength of their naval forces positioned at strategic locations from West Africa on the Atlantic Ocean to Goa and Malacca on the Indian Ocean. However, their naval seapower was desperately overstretched in its attempts to police these seaborne trade routes whilst also trying to combat the persistent incursions of the Ottoman fleet throughout the Red Sea and the Indian Ocean. Even more significantly the Portuguese crown had attempted to establish a complete monopoly on all pepper coming into Europe via the Cape route from 1504, in order to fix prices and thus ensure a guaranteed profit.[1] However, this strategy began to backfire almost as soon as it was instituted by Manuel I, who became known as the 'grocer king'. Dissatisfied local pepper merchants throughout the Indian subcontinent returned to the old overland trade routes to sell their pepper to Venetian merchants, who were prepared to negotiate the price. The decision to establish a spice and pepper monopoly in Lisbon also had serious repercussions for German capital, which had relocated there in the aftermath of da Gama's voyage to ensure that merchants could buy these imports and ship them to markets in northern Europe. In 1516 the commercial representative for the influential House of Fugger, Cristóbal de Haro, withdrew from Lisbon in protest at the new monopoly, which was making it increasingly difficult for the Fugger to reap any substantial profit from the pepper trade.[2] De Haro chose to settle in Seville, where he hoped to encourage Castile to challenge the Portuguese spice and pepper monopoly by following in Columbus' footsteps and sailing westwards, rather than eastwards, to capture the final prized possession in the spice trade: the Moluccas Islands.

The spices which had trickled into the markets of mainland Europe for years were one of the most sought after condiments within the domestic economy of early modern Europe (illus. 38). Not only were they used to flavour monotonous and often rancid meat dishes, but as their reputation grew they became renowned for their various abilities to strengthen eyesight, cure dropsy and eliminate liver pains. To capture the tiny islands of the Moluccas was effectively to capture one of the most plentiful supplies of spices available throughout Southeast Asia. Situated to the east of Malacca on the rim of the Pacific Ocean, the islands had been for many years the primary source of high-quality pepper distributed throughout the trading emporia of the Indian Ocean. In 1511 the Portuguese naval commander d'Albuquerque had captured the commercially vital city of Malacca on the southern tip of the Malay peninsula, to where the majority of the pepper produced in the Moluccas was shipped for wider distribution. Its strategic importance to the Portuguese had been emphasized by the diplomat Tomé Pires, who noted in his *Suma Oriental* written during the same period that:

Men cannot estimate the worth of Malacca, on account of its greatness and profit. Malacca is a city that was made for merchandise, fitter than any other in the world.[3]

From their base in Malacca the Portuguese endeavoured to take possession of the Moluccas and claim the final link in the complicated commercial chain which constituted the spice trade. However, the Portuguese crown came up against a diplomatic obstacle even greater than the navigational difficulties which faced its sailors in reaching the Moluccas. The Treaty of Tordesillas, which had been signed in 1494 in the aftermath of Columbus' discoveries in the western Atlantic, had provisionally settled the territorial dispute between Portugal and Castile as to where their mutual spheres of territorial influence extended. However, by fixing the line of demarcation 370 leagues west of the Cape Verde Islands, in the Atlantic, it only postponed the inevitable diplomatic dispute as to where this line should fall in the eastern hemisphere, if it was to be extended and drawn right round the globe. The treaty was clearly drawn on a flat map of the known world, rather than a globe which would have allowed the two crowns to partition the world like splitting an apple in two. However, geographical knowledge in 1494 could literally not encompass the entire globe as, prior to the Portuguese voyages, nothing was known with any degree of accuracy of the geographical coordinates of the Indian Ocean, or those of the still as yet undiscovered Pacific Ocean. In 1513, as the Portuguese dispatched voyages to raid the still independent Moluccas islands, the Castilian adventurer Vasco Núñez de Balboa crossed the Isthmus of Darien in Central America and became the first known European to sight the Pacific Ocean.[4] News of his sighting fuelled growing specula-

38 Boucicaut Master, 'Pepper harvest in Coilum', illumination from Marco Polo, *Livre des Marveilles*, *c.* 1410. Bibliothèque Nationale de France, Paris.

tion as to where the line of demarcation between the two crowns was to be drawn in the eastern hemisphere, somewhere between Balboa's discovery in the west and the Portuguese in Malacca to the east.

In the rapid attempt to assimilate such information, the apprehension of the world of early modern diplomacy took on a perceptibly global perspective. The significance of Balboa's discovery and the Portuguese pursuit of the Moluccas gave the early modern world a definably global outlook with the prize, to whoever could complete this geographical circle, of virtual control of the lucrative spice trade. This chapter deals with the commercial competition for the Moluccas Islands which preoccupied European diplomacy for nearly two decades, and the remarkable diplomatic and geographical debates which it produced. The political controversy which surrounded the Moluccas placed cartography at the centre of the debate, as both geographers and diplomats strove to accommodate the implications of the Portuguese and Castilian voyages into their world picture, whilst princes and merchants sought to utilize both maps and globes as central bargaining apparatuses in their political dispute over the spice islands. The dispute was to inspire not only the first complete circumnavigation of the globe, but also an explosion in global cartography which established a whole industry of globemaking as the discipline of cartography and a whole range of geographers and mapmakers cashed in on the political and cultural esteem accorded to their intellectual labours. The dispute ultimately established the contours of a recognizably modern global image of the world, as well as finally professionalizing the

discipline of geography as a highly politicized, as well as educationally influential, field of intellectual enquiry.

Sailing West to the East: The Quest for the Moluccas

The news of Balboa's sighting of the Pacific in 1513 soon reached Castile,
and the ears of the Fugger representative who had recently moved to Seville, Cristóbal de Haro. De Haro appears to have been more convinced
than ever of the possibility of following Columbus and Balboa by sailing
west into the Atlantic to take possession of the Moluccas in the 'east'.
Throughout 1516 he began developing plans to finance a voyage from Castile to the islands. By early 1517 he had acquired the services of a pilot
whom he believed had the ability to lead an expedition to the spice islands
by sailing south-west: the Portuguese sailor Fernão de Magalhães, better
known by his Hispanicized name Ferdinand Magellan (illus. 39). Magellan
had already spent several years in the Portuguese fleet in the Indian Ocean,
and had taken part in the capture of Malacca in 1511.[5] He was thus perfectly equipped with the nautical and commercial knowledges required to
launch a renewed assault on the Moluccas. By encouraging him to undertake such a voyage under the Castilian flag, de Haro and his business
colleagues emphasized the extent to which economic considerations over-
rode national affiliation in the commercial pursuit of an alternative trade
route to the islands. The first account of Magellan's now famous voyage
was written by Maximilianus Transylvanus, counsellor to Charles V, a
student of the famous geographical scholar Peter Martyr and relative of
Cristóbal de Haro. Transylvanus published his account of the voyage in
Cologne in January 1523. Entitled *De Moluccis Insulis*, Transylvanus' text
recounted the reasons for de Haro's choice of pilot, and the commercial
motivations which lay behind the projected voyage:

Four years ago, Ferdinand Magellan, a distinguished Portuguese who had for
many years sailed about the Eastern Seas as admiral of the Portuguese fleet,
having quarrelled with his king who he considered had acted ungratefully towards him, and Christopher Haro, brother of my father-in-law, of Lisbon, who
had through his agents for many years carried on trade with those Eastern countries and more recently with the Chinese, so that he was well acquainted with
these matters (he also, having been ill-used by the king of Portugal, had returned to his native country, Castile), pointed out to the emperor [Charles V]
that it was not yet clearly ascertained whether Malacca was within the boundaries of the Portuguese or the Castilians, because hitherto its longitude had not
been definitely known; but it was an undoubted fact that the Great Gulf and the
Chinese nations were within the Castilian limits. They asserted also that it was
absolutely certain that the islands called the Moluccas, in which all sorts of
spices grow, and from which they were brought to Malacca, were contained in

39 *Portrait of Magellan*, engraving in Theodor de Bry, *Voyages*, Frankfurt, 1610. British Library, London.

the Western, or Castilian division, and that it would be possible to sail to them and to bring the spices at less trouble and expense from their native soil to Castile.[6]

Transylvanus' text, which was widely circulated amongst the financial community of Augsburg, makes it quite clear that not only was the singular aim of the voyage to 'search for the islands in which spices grow', but also that de Haro's financial input into the voyage was crucial to the whole project. The influential Castilian historian of the Americas, Bishop Bartolomé de Las Casas, in his account of Magellan's justification for basing his project in Castile, recalled the significance of the geographical information which the Portuguese sailor took with him when he left Lisbon:

Magellan brought with him [to Seville] a well-painted globe showing the entire world, and thereon traced the course he proposed to take, save that the Strait was purposely left so that nobody could anticipate him ... I asked him what route he proposed to take, he replied that he intended to take that of Cape Santa Maria (which we call Rio de la Plata) and thence follow the coast up [south] until he found the Strait. I said, 'What will you do if you find no strait to pass into the other sea?' He replied that if he found none he would follow the course that the Portuguese took. But, according to what an Italian named Pigafetta of Vincenza who went on that voyage of discovery with Magellan, wrote in a letter, Magellan was perfectly certain to find the Strait because he had seen on a nautical chart made by one Martin of Bohemia, a great pilot and cosmographer, in the treasury of the King of Portugal the Strait depicted just as he found it. And, because the said Strait was on the coast of land and sea, within the boundaries of the sovereigns of Castile, he [Magellan] therefore had to move and offer his services to the king of Castile to discover a new route to the said islands of Molucca and the rest.[7]

Magellan's utilization of knowledge apparently gleaned from Martin Behaim's geographical information is supported by the travel account subsequently published by Antonio Pigafetta, who sailed with Magellan. In his account Pigafetta noted that:

The captain-general said that there was another strait which led out, saying that he knew it well and had seen it in a marine chart of the King of Portugal, which a great pilot and sailor named Martin of Bohemia had made.[8]

What is so significant here is that Magellan used Behaim's geographical information not necessarily for its precision, but for its globular dimensions, the only dimensions that would have been of any use to him as he sought to join up the two hemispheres of the world. It is also telling that in the face of the internationalist nature of de Haro's assault on the Moluccas, allegiance to crown and country were rapidly sacrificed. The commentaries of Transylvanus and Las Casas make it clear that this was a commercial venture which seamlessly transgressed established limits of national and imperial sovereignty. Magellan's projected voyage was in fact a trading venture which brought together a Portuguese pilot (Magellan), using a chart (or globe) belonging to a German (Behaim), employing an Italian chronicler (Pigafetta) in a German-financed voyage, cashing in on his privileged access to sensitive Portuguese information, and flying under the flag of Castile. The level of diplomatic, technical and intellectual expertise required for such an ambitious long-distance commercial voyage ensured that money was quickly made available to purchase such knowledge, regardless of the obstacles of national or regional allegiance.

Nor was Magellan the only specialist lured to Seville by de Haro. By the time preparations for the voyage were complete in the summer of 1519, the team of geographical experts and advisers who surrounded Magellan included Ruy Faleiro, appointed chief pilot and supervisor of maps and instruments, the brothers Pedro and Jorge Reinel, cosmographical advisers, and Diogo Ribeiro, official chartmaker to the voyage.[9] All four of these highly respected figures were not only Portuguese, but had also all been appointed by royal command to positions of authority in the

Portuguese *Casa da Índia*, which oversaw the commercial and geographical regulation of trade along the seaborne route to India. In another account of the motives behind Magellan's decision to sail south-west to the Moluccas, the Castilian historian Bartholomé Leonardo de Argensola, in his *Conquista de las islas Malucas* published in Madrid in 1609, outlined the involvement of Magellan's Portuguese compatriots in the voyage, as well as offering further specific details on his intellectual and geographical justification for the voyage. Argensola recorded that Magellan, returning from his stint in Malacca in 1511:

> Consider'd, that since the *Moluccos* were 600 Leagues East from *Malacca*, which make 30 Degrees, little more or less, they were out of the *Portuguese* Limits, according to the antient Sea Chart. Returning to *Portugal* he found no Favour, but thought himself wrong'd, and resenting it, went away into *Castile*, carrying with him a Planisphere, drawn by *Peter Reynel*; by which, and the Correspondence he had held with *Serrano*, he perswaded the Emperor, *Charles V* that the *Molucco* Island belong'd to him. It is reported, That he confirm'd his opinion with Writings, and the authority of *Ruy Faleyro*, a *Portuguese* Judiciary Astrologer.[10]

Argensola's account of Magellan's reasons for defecting from Portugal to Castile suggests that the Portuguese sailor was convinced that if the principles of the Treaty of Tordesillas were applied to the geographical position of the Moluccas the islands would, in effect, lie within the 180 degrees of the globe allocated to the Castilian crown and its regent, the Habsburg emperor Charles V. His decision to undertake such a hazardous voyage was therefore made after extensive practical and intellectual consultation of the widest possible variety of geographical information available, including not only maps, charts and globes, but also the more esoteric cosmographical knowledges of savants like Faleiro. His reliance on such learned geographical speculation was underlined by the mass of globes, instruments and charts with which he fitted out his fleet. As well as 'two planispheres belonging to Magellan and made by Pedro Reinel, and other charts for the navigation to India',[11] the remnants of the fleet which returned to Europe in 1522 still retained an impressive array of navigational equipment and geographical artefacts, which included:

> 23 charts by Nuño García, an Italian cartographer employed by the Council of the Indies; 6 pair of dividers, 7 astrolabes (one of brass), 21 wooden quadrants such as Columbus used, 35 magnetized needles for the compasses, and 18 half-hour glasses for keeping time . . . [and] two planispheres by Pedro Reinel.[12]

Portuguese agents based in Seville kept their king continually informed on the progress of the preparations for the voyage, and the importance of the maps and charts of the Reinel brothers and Diogo Ribeiro to the whole enterprise. On the eve of Magellan's departure in the autumn of 1519 one of João's agents reported to him with further details of the forthcoming voyage:

The route which is said to be followed is from San Lucar directly to Cape Frio, leaving Brazil to the right side until after the demarcation line and from there to sail West ¼ North West directly to Maluco, which Maluco I have seen represented on the round chart made here by the son of Reinel; this was not finished when his father came there, and his father achieved the whole thing and placed the Molucas lands. From this model have been made all the charts of Diego Ribeiro and also the particular charts and globes.[13]

Magellan's navigational plans were clearly no longer informed simply by the logistical difficulty of finding his way across open seas to the Moluccas. As the comment on sailing either side of 'the demarcation line' makes clear, the artificially created lines of territorial division established by the Iberian crowns in the Treaty of Tordesillas attempted to circumscribe the movement of sailors and pilots, whose world view became increasingly defined through the decisions made by diplomats and merchants back in Europe.

By September 1522 preparations for the voyage were complete, and on 22 September Magellan's fleet of five ships sailed out of the Castilian port of San Lúcar de Barrameda bound for the Moluccas. Sailing out into the Atlantic, Magellan proceeded down the coast of Brazil and then sailed southwards in an attempt to find a strait which would allow him passage into the Pacific Ocean. By October 1520, after stopping several times to reassess his position, he found his way into the strait which now bears his name. By the end of November 1520, after much cautious searching, Magellan had emerged from the strait into the open sea of the Pacific Ocean. From there he set out across the Pacific. One of the difficulties which subsequently confronted him was that he was sailing in virtually uncharted waters, relying on the conceptual geography of the *Geographia* to guide him. As Ptolemy's calculations had substantially underestimated the extent of the oceans east of India, the journey was far more arduous than either Magellan or any of his pilots had envisaged. By the spring of 1521 the exhausted fleet arrived in the eastern Philippines, where Magellan himself was killed on 27 April after a skirmish with locals on the island of Mactan. The remnants of the fleet sailed on towards their destination, and on 6 November they sighted the islands of the Moluccas. After loading a substantial cargo of spices, pepper and cloves, they set sail for Castile via the Cape of Good Hope, led by Juan Sebastián de Elcano. On 8 September 1522 the remaining 18 men of an original crew of 240 arrived back in Seville, having completed the first recorded circumnavigation of the globe.

The successful return of the remnants of Magellan's fleet caused a diplomatic sensation. The Portuguese crown, upon hearing of its return, immediately lodged a protest with the sovereign of Castile, the recently appointed Holy Roman Emperor Charles V. The Portuguese claimed that Magellan had effectively encroached upon territory which they claimed under the terms of the Treaty of Tordesillas. However, the implications of Magellan's voyage went far beyond this sudden renewal of the long-standing territorial conflict between the two Iberian crowns. News that Magellan's fleet had circumnavigated the globe was received by a diplomatic community throughout mainland Europe which was gradually beginning to accept the social and political effects of long-distance commercial travel. The Imperial Diet of Nuremberg, which met in 1522, received the news of the voyage in the midst of its attempts to come to terms with not only the rise of Lutheranism throughout northern Europe, but also the difficulties of forming a united Christian front against the Ottoman empire. Perhaps even more significantly, it was locked in debate over the ethical and political problems stemming from the recent growth of overseas trade and commercial monopolization. Concerned at the amount of precious metal leaving their territories to pay for pepper bought in the Indian Ocean, the German princes present at the Diet inveighed against the growing political influence of powerful merchant families such as the Fugger and the Welser, and their hold over the movement of spices between Lisbon and Antwerp which allowed them to dictate the (relatively high) price of spices and pepper. The merchants responded by pointing to the prosperity of the southern German states, and the potentially disastrous economic consequences for them if the monopolies were lifted.[14]

News of the return of Magellan's expedition could only strengthen the hand of the German merchant houses, who stood to profit enormously from a voyage which they had effectively financed. Information on the voyage was also eagerly anticipated by prominent supporters of Charles V, who had been persuaded by the German merchants to support the voyage. They included one of the most senior delegates at the Diet, Francesco Chiericati, the papal nuncio to Germany. Chiericati regularly wrote to Isabella d'Este Gonzago in Mantua, telling her of the Diet's activities and resolutions. Throughout December 1522 he informed Isabella of his attempts to persuade the German states to participate in a Hungarian-Croatian front against the Ottomans, whilst also supplying her with his own sensational news of Magellan's voyage:

My servant from Vincenza [Pigafetta] whom I sent from Spain to the Indies has returned highly enriched with the greatest and most wonderful things in the world, and has brought an Itinerary from the day he left Spain until the day of his return – which is a wonderful thing. Your Most Illustrious Ladyship shall soon be acquainted with everything.[15]

On 10 January 1523, writing of the growing influence of the 'heretical' ideas of one Martin Luther, Chiericati wrote to Isabella enclosing substantial sections of Transylvanus' account of Magellan's voyage:

First of all, they sailed southwards to those islands in the Oceanic Sea which are called Terra Ferma, and round the point, over the Sea of Sur towards the west. Then, turning to the north and east, they found themselves in the great gulf, near the Spice Islands, and sailed through the Persian and Arabian Seas, by the Cape of Good Hope, into the Ethiopian Sea and across the Atlantic, until they reached the Canary Islands, and returned to their own land by the opposite way, having gained not only great riches, but what is worth more – an immortal reputation. For surely this has thrown all the deeds of the Argonauts into the shade. Here we have read a long account of the expedition, which the Emperor Charles has sent to the Archduke Ferdinand, who has kindly shown it to me, and has also given me some of the spices which they brought back from those parts, with boughs and leaves of the trees from which they came. Charles has also sent the Archduke a painted map of the journey, and a bird which is very beautiful, which the kings of those countries bear with them when they go into battle, and say they cannot die so long as it is at their side.[16]

Chiericati's account of the increased diplomatic (and implicitly commercial) significance of the geographical artefact produced as a result of the voyage, the 'painted map', was symptomatic of the centrality of cartography in defining the discursive contexts within which it became possible to discuss the issue of the Moluccas. Charles' presentation of the map to his brother the Austrian archduke Ferdinand shares similarities with the ritual presentation of maps from faraway lands carried out by Portuguese sailors on their return to Lisbon. However, the commercial significance of Magellan's voyage was even more directly tangible than the early tentative Portuguese voyages into the Indian Ocean. Charles' map was displayed alongside the spices and exotic birds recovered from the journey. This should not be perceived as simply the ostentatious display of exotica but rather, as the presence of the map suggests, a nascent classificatory procedure which attempted to control the knowledges and commodities produced by long-distance travel. The map offered a fantasy of territorial control over the Moluccas where none in fact existed. However, it also laid itself open to extensive manipulation by political and commercial interests regardless of its geographical accuracy. Within the specific context of Chiericati's relation of diplomatic intrigue to d'Este, the map exuded a particularly resonant message to Ferdinand, exhibiting Charles V's control not only over the production and movement of such materially luxurious objects as world maps, but also over the new dominions depicted on the globe which flew the flag of Castile. Ferdinand was therefore asked not only to accept the territorial gains made by the likes of Magellan on behalf of Charles, but also to consider the commercially successful acquisition of prized merchandise such as the spices presented alongside the world map. This display of intellectual expertise and political rhetoric, packaged into one geographical object, united Charles' claims to both territorial possession and commercial authority, claims which were to symbolize the clash between Charles and João III over the subsequent seven years of protracted diplomatic debate over the possession of the Moluccas.

This calculated display of the possession of cartographic 'information' also characterized the pages of the sketchbooks kept by Pigafetta whilst travelling with Magellan. In his sketch map commemorating the arrival of the fleet in the Moluccas, he depicted the islands of '*Malluco*' above a drawing of a clove tree labelled '*arbore di garofali*' (illus. 40). His shrewd depiction of the islands emphasized their commercial potentiality, in comparison to which the urgency of recording their hydrographic position and bearing came a poor second. As Pigafetta's visual representation of the Moluccas and the commercial impetus behind the circulation of the maps of Ribeiro and the Reinels indicate, such maps laid claim to possession primarily within a commercial, rather than a national, context. However, even this situation precludes a simplistic reading of the systematic implementation of colonial domination. One of the most effective aspects of the possession of a map recovered from the rigours of long-distance travel was its ability to instil in its owner a belief in his possession of the territory depicted, or, within the concerns of commercial entrepreneurs, access to the territories portrayed in their maps and across their globes. Such beliefs were invariably reinforced by the producers of maps and charts (such as Behaim and Berlinghieri) in their attempts to establish favourable relations with a particular patron. However, the belief in effective possession of cartographic territory was an illusion, and nowhere more so than in the case of the Moluccas, whatever someone like Pigafetta may have claimed upon his return to Europe. His maps should therefore be seen as defining a provisional field of economic exploration, rather than the coherent implementation of a regime of colonial expansion. Such fantasies may have been entertained across the negotiating tables back in Lisbon, Seville and Saragossa, but their cartographic representations remained just that: fantasy projections of possession.

Because of their intense association with the precious merchandise which they had in effect helped to obtain, maps and globes of the type pre-

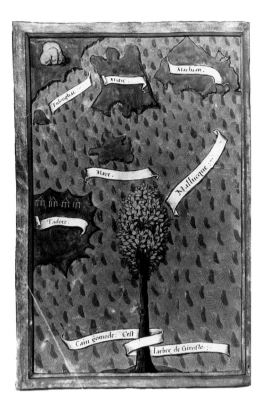

40 Antonio Pigafetta, chart of the Moluccas, *c.* 1522, vellum. Yale University Library, New Haven.

sented to Ferdinand by Charles V became imbued with the commercial ethos of the enterprise as a whole. The geographical paraphernalia which Magellan took with him was itself a calculated investment, a 'capital good' taking its place alongside the other goods which were shipped by de Haro and which passed for commodities. The inventory for the voyage included objects for trade such as 20,000 hawks bells, 500 pounds of glass beads, 400 pairs of German knives, a thousand mirrors (graded according to quality), brass bracelets, fish hooks, silk and cotton (of various colours), and silk robes 'made in the Turkish fashion'.[17] It was precisely these objects, alongside dividers, quadrants, astrolabes and, perhaps most significantly, maps and globes, which were displayed, brandished and offered in gift exchanges as Magellan moved throughout the Pacific.

One particularly significant incident which highlights the ways in which geographical knowledge was deployed beyond the limits of its cultural production in an attempt to inspire awe and wonder, was described by Pigafetta in his chronicle of the voyage. He recorded that in March 1520 Magellan stopped for refreshment at Limasawa in the Philippines, after successfully finding his way into the Pacific Ocean. Pigafetta noted that in his first meeting with the local king:

The captain gave this king a robe of red and yellow cloth, made in the Turkish fashion, and a very fine red cap ... After that the captain showed him cloths of divers colours, and linen, coral, and many other goods, and all the artillery, several pieces of which he caused to be fired before his eyes, whereby the king was greatly astonished.[18]

The representation of the 'astonished' response of the king is repeated in his reaction to Magellan's next gesture, which was to show him:

A great number of swords, of cuirasses, and bucklers, and then he made two of his men exercise at swordplay before the king. And he showed them the marine chart and the compass of his ship, telling him how he had found the strait [by which] to come hither, and of the time which he had spent in coming, also how he had not seen any land. At which the king marvelled.[19]

In this account, the map is not only coextensive with the militarism of the sword and the cannon, but it also partakes of the luxury and preciousness of the 'cloths of divers colours' which Magellan exhibits to the king. As with earlier Portuguese attempts to ratify the power of the map to those unfamiliar with its significance, the encounter foregrounds Magellan's desire to stress the technical superiority of the European map. The discovery of the passage from the Atlantic to the Pacific, the enormous distances involved, the amount of time spent travelling and the navigational skill required to sustain a coherent course out of sight of land, are all treasured elements of Magellan's compasses and charts, but have no meaning whatsoever for the king. Magellan's comments to him are again a reflection on his own admiration and wonder at the advantages of the chart and the globe, rather than an accurate account of the ruler's amazement. The map negotiated an anxious point of encounter for Magellan, displayed as an object encapsulating both technical superiority (underlined by the concomitant display of weaponry and claims to navigational expertise) and ostentatious wealth (as it is compared to the fine cloth and coral carried on board as precious cargo).

It was on their return from these particularly fraught and contested encounters overseas that the maps and charts taken on the voyage, and those created en route, came to be re-established by their patrons and backers as particularly precious objects to be displayed as concrete instantiations of the voyages made, and the commercial objectives achieved. The surviving commander of the voyage, Juan Sebastián de Elcano, was clearly aware of this when making his report to Charles V in 1522 on its outcome. Elcano was the most senior member of Magellan's expedition to survive, and in his report on its achievements, he wrote that:

We discovered many very rich islands, among them Banda, where ginger and the nutmeg grow, and Zabba, where pepper grows, and Timor, where sandalwood grows, and in all the aforesaid islands there is an infinite amount of ginger. The proof of all these productions gathered in the same islands in which they grow, we bring to display to Your Majesty.[20]

Elcano's account offers no doubt as to the aim of the voyage. However, within his confident elaboration of the goods to be found in the islands is a significant moment of ambiguity. Is the 'proof' the products themselves shipped back and carefully recorded, or is it the maps, charts and globes brought back from the voyage? Such objects were approached with veneration and increased admiration, as being imbued with an aura of esoteric knowledge as a direct result of having survived the rigours and dangers of long-distance travel. The maps brought back by the survivors of the expedition were consequently viewed as precious goods, prized possessions handled along with the cargo. These maps were themselves as eagerly valued as cloves and spices, and because of their movement and circulation took on an even greater economic and political significance. The terrestrial globe depicted in Hans Holbein's famous painting *The Ambassadors*, painted in 1533, was based on an original Nuremberg globe which meticulously traced Magellan's voyage across its surface (illus. 41). Central to the composition's complex representation of diplomatic exchange and political authority, the position of the terrestrial globe within the painting is symptomatic of the extent to which, in the aftermath of the voyage, geographical knowledge became increasingly vital in the establishment of political claims to territorial possession and commercial expansion.

In response to Portugal's claim that Magellan's voyage violated Portuguese sovereignty in Far Eastern waters, Charles V cleverly offered to submit the dispute to diplomatic arbitration.[21] Whilst the Portuguese were eager to comply, believing that their superior navigational knowledge of the area would confirm their right to the Moluccas, their decision even to countenance arbitration was fatal, as it was tantamount to admitting that Charles had some claim on the spice islands. The first meeting of the two crowns to discuss the issue took place at Vittoria in February 1524, where João III and Charles V agreed to debate 'ownership, possession ... navigation and trade'.[22] Whilst initial negotiations proved fruitless in establishing possession, it did lead to the two crowns agreeing in principle that 'neither of the parties to this treaty shall dispatch expeditions to Molucca for purposes of trade or barter',[23] and that a working party should be appointed to study the controversy on both sides, and reconvene at a later date along the border of Castile and Portugal, between Badajoz and Elvas. In the meantime, Charles wasted no time in compiling a diplomatically and geographically compelling case for possession of the Moluccas.

By the time the two crowns sat down to discuss the issue again at

41 Hans Holbein, *The Ambassadors*, 1533, oak. National Gallery, London.

Badajoz-Elvas in April 1524, Charles had attempted to reconstruct the retinue of predominantly Portuguese geographers who had helped Magellan
in preparation for his voyage five years earlier. The delegations consisted of
nine members on each side, as well as technical advisers, mostly comprised
of geographical experts. On the Castilian side sat the Portuguese cartographer Diogo Ribeiro, whose charts had been so influential in planning the
route taken by Magellan in 1519. However, by the time the delegates sat
down at Badajoz-Elvas the Reinel brothers, Pedro and Jorge, who were so
central to Magellan's original navigational plans, were now to be found
amongst the Portuguese delegation. Losing the brothers was clearly a
blow to the Castilian team who had, as the Portuguese delegate Lopez de
Siqueira reported in a letter to João III, attempted to woo the Reinel
brothers back into the emperor's camp:

I stayed at home these last two days, during this time Pedro Reinel came to see
me and told me confidentially that he has been invited together with his son to
enter the Emperor's service. The latter has addressed to them letters signed by
himself and the same proposal was made to Simão Fernandez.[24]

Both the Reinel brothers and Fernandez, a geographical adviser to the Portuguese negotiators, were offered employment with the Castilian delegation at salaries of 30,000 *reis* a year, more than double the average salary offered to such specialists in the pay of the *Casa da Índia* in Lisbon. Charles' offer was clearly an attempt to ensure a monopoly on the most advanced geographical information available, as well as an unscrupulous endeavour aimed at neutralizing any Portuguese information which would have possibly confirmed Portugal's claim to the Moluccas.

However, the Castilian negotiating team had managed to retain the services of Ribeiro, as Richard Eden makes clear in his account of the talks contained in his translation of Peter Martyr's *Decades of the New World*, published in English in 1555:

Nunno Garcia, Diego Rivero, being all expert pilots & cunning in making cardes for the sea, shuld be present, & brynge foorth theyr globes and mappes with other instrumentes necessarie to declare the situation of the Islandes of the Malucas abowt which was al the contention and stryfe.[25]

Ribeiro's career, from his involvement in Magellan's voyage to his participation in the diplomatic dispute over the position of the Moluccas, became symptomatic of the increasing political value and subsequent professionalization accorded to the sixteenth-century geographer. On the eve of Magellan's expedition in March 1519 he had been paid four gold ducats by the Castilian crown for the preparation of four astrolabes, and by May he had received eight silver *reals* for charts which he had prepared, presumably for the forthcoming voyage.[26] When the two crowns sat down to discuss the Moluccas in 1524 Ribeiro was well established in the pay of the Castilian crown. Living in Coruña, he was employed as the official cartographer to the Castilian *Casa de la Especieria* under the direction of Cristóbal de Haro, the Fugger agent responsible for initially backing Magellan's voyage. As well as making sailing charts, spheres, world maps and astrolabes, Ribeiro was busy designing a metal water pump which, if successful, was to double his wages from 30,000 to 60,000 *maravedis*. By 1527 de Haro was commissioning Ribeiro to provide Castilian pilots with world charts to enable them to navigate their way to the spice islands by way of the Cape of Good Hope as well as the newly discovered route through Magellan's Strait.[27] Exhibiting his geographical knowledges of the Cape route gleaned from time spent in the employ of the Portuguese crown in Lisbon, as well as his experience gained plotting Magellan's course to the Moluccas via the west, Ribeiro's expertise was clearly of invaluable help to the Castilian team at Badajoz-Elvas in trying to ratify their claim to the spice islands.

The obsession with cartographical information and the politics of geographical partition which preoccupied the delegates at Badajoz-Elvas was amusingly underlined by Richard Eden's account of the arrival of the Portuguese negotiating team at the small border town of Quadiana:

It so chanced that as Frances de Melo, Diego Lopes of Sequeyra, and other of those Portugales of this assemble, walked up the river side of Quadiana, a little boy who stood keeping his mothers clothes which she had washed, demanded of them whether they were those men that parted the world with The emperour. And as they answered, yea: he tooke up his shirt and shewed them his bare arse, saying: Come and drawe yowre line here throughe the middest. Which saying was afterwarde in every mans mouthe and laughed at in the town of Badajoz.[28]

The line was in fact not conclusively drawn anywhere at Badajoz-Elvas. Having failed to retain the services of the Reinel brothers, the Castilian delegates were apparently unable to present a geographically conclusive case for possession of the Moluccas. Part of the disputation revolved around an initial inability to redefine the exact line of demarcation between the two crowns that had been established under the Treaty of Tordesillas. By insisting that the line in the west be fixed on the island of Sal in the Cape Verdes, the Portuguese claimed that the Moluccas lay 137 degrees east of this point, whilst the Castilian delegation argued that the islands lay 183 degrees east of Sal (an astonishing difference of opinion) which, if correct, would place them just three degrees within the Castilian sphere.[29] As neither side possessed the scientific information required to establish a precise calculation for the measurement of longitude, such differences of opinion could never be fully resolved and the talks reached stalemate.

Charles V attempted to take advantage of the diplomatic impasse over the next two years by sending a series of fleets along Magellan's route to take the Moluccas by force. However, they consistently failed to successfully negotiate the enormous distances involved in sailing from Seville to the Moluccas via Magellan's Strait, and in particular the formidable difficulties involved in crossing the still relatively uncharted Pacific Ocean. Such voyages were in fact political folly, as the Castilian crown persistently underplayed the length of the ocean: to admit to its real size would have been tantamount to admitting that the Moluccas fell within the Portuguese half of the globe. By the beginning of 1529 Charles had realized the futility of pursuing an increasingly unprofitable claim to the islands, and began to develop another strategy to extract some level of financial gain from his claim to them. One potential solution which he had initially pursued was to combine negotiations over the Moluccas with the arrangements for João's marriage to Charles' sister Catherine, which took place in 1525. The emperor had initially offered to pay the Portuguese king a dowry of 200,000 ducats, but, as a result of the dispute over the Moluccas in 1524, then had the nerve to propose to João that the amount be waived, at the rate of 40,000 ducats per year, in return for unlimited Portuguese access to the Moluccas for just six years, after which possession would revert to Charles![30] It is no wonder that in the midst of these diplomatic and matrimonial machinations one of the most celebrated artistic manifestations of the union between João and Catherine should have been the *Spheres* tapestry series discussed in Chapter One, which prominently displayed Portuguese flags flying over the contested territories of the Indonesian archipelago.

By the spring of 1529 Charles had agreed with João to return to the negotiating table in an attempt to come to some financial arrangement over the vexed issue of the possession of the Moluccas. Charles was in desperate need of money in the face of his wars with the French sovereign François I, and João was keen to safeguard his overseas commercial interests without having to go to war with his powerful brother-in-law. As a result diplomatic delegations again convened a meeting, this time at Saragossa, to agree the terms of a new and binding treaty. The weight of geographical evidence in favour of Charles' claim to the islands was once more invoked, but this time in the form of political rhetoric designed to strengthen any financial settlement which he felt it advantageous to accept. On 23 April 1529 the two crowns signed a treaty which effectively put an end to the controversy concerning possession of the Moluccas. At first glance the treaty appeared to signal defeat for the emperor. Charles gave up his claim to the islands in return for which the Portuguese crown had to pay him compensation of 350,000 gold ducats. Charles also negotiated a clause in the treaty which allowed him to reclaim his right to the islands if new geographical evidence emerged, upon return of the 350,000 ducats. A line of demarcation was to be drawn 17 degrees east of the islands on a standard map which placed them firmly within Portugal's sphere of influence. Any Castilians found infringing the new line could be punished by the Portuguese.

Once more, it was the figure of the map that stood at the centre of the settlement reached over the Moluccas, a contractual document binding the two sides in diplomatic agreement. The wording of the draft treaty first drawn up on 17 April emphasized the extent to which the standardization of the figure of the map was crucial in ascertaining the line of demarcation between the two sides and establishing diplomatic and commercial concord:

In order that it may be known where the said line falls, a model map shall at once be made on which the said line shall be drawn in the manner aforesaid, and it will thus be agreed to as a declaration of the point and place through which the line passes. This map shall be signed by the said lord Emperor and King of Castile, and by the said lord King of Portugal, and sealed with their seals. In the same manner, and in accordance with the said model map, the said line shall be drawn on all the navigation charts whereby the subjects and natives of the kingdoms ... shall navigate. In order to make the said model map, three persons shall be named by each of the said lord kings to make the said map upon oath, and they shall make the said line in conformity to what has been said above. When the map has thus been made, the said lord Emperor and King of Castile and the said lord and King of Portugal shall sign it with their names, and shall order it to be sealed with the seals of their arms; and the said marine charts shall be made from it as aforesaid, in order that the subjects and natives of the said lord kings may navigate by them so long as the said lord King of Castile shall not redeem and buy back the said right.[31]

The treaty was thus a remarkable political document in which the ritual construction of the map, and the fantasy of its ability to implement a form of maritime policing (that was in reality unenforceable), stood as the object through which the diplomatic transactions which carved up the Moluccas came to be ratified. Having been established as a 'capital good' which entered the market-place alongside the spices and cloves it projected, the map now appeared in the more restricted arena of the political economy of diplomatic negotiation. It was this politically sensitive arena which appreciated the map as a type of visual contract. The projected map which was to emerge from the Treaty of Saragossa was to be purged of the more openly acquisitive spirit of Pigafetta's sketch map, striving to translate the commercial possibilities opened up by such mapmakers into the political authority and power which accrued to the empires who laid claim to the spice-producing islands of the Moluccas.

However, it should be stressed that the maps produced both before and immediately after the settlement reached at Saragossa were by no means impartial. The agreement ratified in April 1529, which fixed the line of demarcation 17 degrees east of the Moluccas, was an official resolution to an issue which the treaty itself nevertheless still conceded was a cartographic compromise. The agreed text of the ratified treaty, in distinction to the initial draft, insisted that:

In order to ascertain where the said line should be drawn, two charts of the same tenor will be made, conformable to the chart in the India House of Trade at Seville, and by which the fleets, vassals, and subjects of the said Emperor and King of Castile navigate. Within thirty days from the date of this contract two persons shall be appointed by each side to examine the aforesaid chart and make the two copies aforesaid conformable to it ... This chart shall also designate the spot in which the said vassals of the said Emperor and King of Castile shall situate and locate Molucca, *which during the time of this contract shall be regarded as situated in such place*.[32]

The standardized map which was to emerge from this treaty clearly placed the Moluccas under Portuguese jurisdiction. However, Charles' insistence on including a clause which allowed him the opportunity to reclaim the Moluccas in the light of any future geographical information underlined the provisional nature of any line of territorial demarcation. The treaty therefore made it quite clear that under different political and commercial conditions the islands could easily be placed in a significantly different geographical position, such as the one which had been provided for them by the Castilian delegation. What shaped the contours of this projected world map, perhaps the most ambitiously planned one to be officially sanctioned by both crowns, was not in fact geographical accuracy but commercial and financial expediency.

There can be no doubt that, despite ceding the rights to the Moluccas, Charles had obtained a highly satisfactory diplomatic and financial result from the treaty. His original claim to the islands in the aftermath of Magellan's voyage was tenuous to say the least. However, one of the most effective bargaining counters which he possessed throughout the negotiations was geographical knowledge, compellingly encapsulated in the maps and charts which were repeatedly displayed and discussed throughout the protracted diplomatic dispute. The awe and respect traditionally accorded to such knowledge was ruthlessly utilized by Charles and his diplomats in the pursuit of diplomatic pre-eminence and financial gain. The treaty enhanced the status of maps and charts as objects valued for their relation to commercial exploitation, as well as establishing geography as one of the most politically valuable domains of intellectual enquiry. The ability of maps to combine compelling geographical evidence of the location of contested territory with a capacity to concretize negotiated political settlement ensured that the discipline of geography was to become enshrined within the intellectual and political life of the sixteenth-century world. Nowhere was this situation more in evidence than in the specific claims to geographical authority made by the specialists in attendance at the talks between Portugal and Castile throughout the 1520s.

Cashing in on the Classics: The Geographical Debate over the Moluccas

The central problem which confronted the Castilian and Portuguese negotiating teams in establishing their claims to geographical accuracy was an inability to base their calculations on the position of the Moluccas upon a commonly accepted body of evidence. As was to be expected under such circumstances, the first authority to which the two sides turned in an attempt to resolve their geographical differences was Ptolemy. However, even his texts, and in particular the *Geographia*, became the site of controversy as the talks developed. This was because it soon became clear that the calculations which Ptolemy developed in completing his work ironically favoured the Castilian, rather than the Portuguese, claims to the Moluccas.

According to Ptolemy the earth had a circumference of approximately 180,000 *stadia* (with ten *stadia* equalling 1.6 kilometres, or 1 mile).[33] Although this estimation greatly reduced earlier calculations produced by Greek mathematicians such as Eratosthenes, it nevertheless still extended the actual figure of the circumference of the earth by one-sixth. Another consequence of Ptolemy's calculations, which took their privileged geographical point of departure as the Mediterranean, was that the extent of the known world, the *oikoumene*, covered over 180 degrees from the Canary Islands to the borders of China. This produced an enormous overestimation of the eastward extension of Asia.[34] The implications of these calculations within the context of the late fifteenth- and early sixteenth-century European voyages of discovery were profound. As a result Columbus had based his westward voyage towards what he perceived to be 'the east' on Ptolemy's arguments, believing that the eastward extension of Ptolemy's projected globe left an extremely short journey to reach China travelling west.[35]

The most influential geographical manifestation of the ways in which these apparently academic issues impinged upon the thinking behind both Portuguese and Castilian plans for maritime expansion was Martin Behaim's terrestrial globe, produced on the eve of Columbus' first voyage. The information contained on it has already been cited as a primary source adopted by Magellan to justify his westward voyage to the Moluccas. It was hardly surprising. Behaim's globe depicts the space between the west coast of Portugal and the east coast of China sailing westwards as occupying a mere 130 degrees. The commonly agreed figure has now been fixed at a far more substantial 230 degrees.[36] Whilst incorporating the Portuguese voyages of discovery up to the point of Dias' rounding of the Cape of Good Hope in 1488, Behaim's globe still retained a predominantly Ptolemaic representation of Asia, which depicted a completely erroneous peninsula, *Sinus Magnus*, to the east of the Malaysian peninsula, thus further overestimating the extent of the eastward trend of Asia. As editions of Ptolemy struggled to incorporate the discoveries of Columbus to the west, whilst still reproducing the accepted belief in the eastward extension of Asia, the as yet uncharted Pacific Ocean was increasingly squeezed into a narrow band of cartographic space. By 1507 Martin Waldseemüller's highly influential world map, in its attempt to reconcile the western discoveries with Ptolemy's depiction of Asia, left the Pacific a breadth of only 70 degrees.[37] Such increasingly distorted representations of the territories to the east of India emerged from the growing intellectual conflict between the academically enshrined authority of the Ptolemaic text and the practical discoveries of the Portuguese geographers and pilots as they sailed deeper into the Indian Ocean throughout the early years of the sixteenth century. As a result, it became increasingly possible to manipulate geographical and navigational data on the size and extent of the territories in question, particularly when substantial financial gain was at stake.

In 1518, as news of Magellan's imminent voyage to the politically sensitive Moluccas began to circulate throughout Seville, Martin Fernandez de Encisco, a former colonial administrator in Española, published his *Suma de Geographie*, an attempt to provide the first authoritative account of world geography to be published in Castile.[38] Dedicated to Charles V, the text borrowed freely from Portuguese navigating manuals, which had by this time standardized their own information on sailing directions and distances. However, in one important instance Encisco's text departed from its reliance on these manuals. The debate over the accurate length of a degree had preoccupied Portuguese geographers and pilots for many years, as they sought to establish an accurate measurement of longitude. After much disputation most commentators had agreed upon a degree as being 17½ leagues. However, in the publication of his *Suma* Encisco departed from the manuals in only one respect: the calculation of the degree. The political nature of his decision to alter its length is strikingly borne out by the history of the printing of his text:

A block was prepared for Encisco's book from the Portuguese original . . . the figure 7 occurring in the value 17½ leagues, which stood against the N-S and S-N rhumbs, has been scraped out (not quite cleanly) and the figure 6 substituted. The rest of the diagram has been allowed to stand, in spite of its inconsistency.[39]

In making this apparently minor alteration Encisco's calculations reduced the Portuguese half-sphere to only 3,000 leagues instead of 3,150. The result was to position the contested Moluccas firmly within the Castilian half of the globe. Whilst Encisco's text did not explicitly foreground his support for Castile's claim to the islands, the importance of his apparently objective mathematical calculation to the subsequent Castilian claim to the Moluccas was obvious.

The politically calculated nature of Encisco's navigational mathematics stressed the extent to which navigational and geographical speculation could be of crucial importance in settling political claims not only to the Moluccas, but also to a range of commercially and territorially contested locations. This was not lost on Charles V and his advisers. In a remarkable statement in his *Conquista de las islas Malucas*, Argensola stressed the impact of this realization upon Charles in his attempt to lay claim to the Moluccas. Argensola noted that the emperor:

Urg'd, that by Mathematical Demonstration, and the Judgement of Men learned in that Faculty, it appear'd, that the *Moluccos* were within the Limits of *Castile*, as were all others, as far as *Malacca*, and even beyond it. That it was no easy Undertaking for *Portugal* to go about to disprove the Writings of so many Cosmographers, and such able Mariners; and particularly the Opinion of *Magellan*, who was himself a *Portuguese*. And that in Case he might be thought partial, because of his being disoblig'd in *Portugal*, that exception did not lie against *Francis Serrano*, who was also a *Portuguese*, and had been favour'd and cherish'd. That to say the Sea Charts had been maliciously contriv'd, was a groundless Objection, and not probable. Besides that, in Relation to the Article of Possession, on which the Controversy depended, it was only requisite to stand by what was writ by, and receiv'd among Cosmographers.[40]

Argensola's comments reveal the extent to which the manipulation of classical learning and geographical knowledge were central to the machinations which enabled Charles to emerge from the Moluccas dispute with such diplomatic and financial credit. Argensola's argument emphasizes the extent to which the learned tradition of geographical enquiry was established for political purposes as the ultimate arbiter of disputation. As far as both he and Charles were concerned, 'it was only requisite to stand by what was writ'. This was clearly a reference to the appreciation that, if the dispute over the Moluccas was referred to the classical authority of Ptolemaic calculations, the Castilian claim to the Moluccas would actually be geographically justified. Argensola stressed that Portugal's claim could only be upheld if it could 'disprove the Writings of so many Cosmographers', indicative of the dilemma in which its cartographers found themselves. Since rounding the Cape in 1488, almost every navigational discovery made by the Portuguese refuted some aspect of Ptolemy's *Geographia*. As such, learned geographical opinion, and particularly opinion based primarily upon his calculations, was very much against their claim to the Moluccas as lying within the Portuguese half of the globe. It is not surprising to learn that, by the time the Castilian delegation met at Badajoz-Elvas in April 1524, their geographical advisers had built up a substantial claim to the islands based on the evidence of Ptolemy. In his opening remarks to the conference, the Castilian delegate Don Hernando Colon insisted in his response to Portuguese claims to the Moluccas that:

... the aid of the old authors, Ptolemaeus and Plinius, is invoked to prove more accurately that the distance was shortened by the Portuguese.[41]

The joint statement of the learned geographical Castilian triumvirate of Tomas Duran, Sebastian Cabot and Juan Vespucci subsequently argued that:

... the description and figure of Ptolemaeus and the description and model found recently by those who came from the spice regions are alike ... Therefore Zamatra, Malacca and the Malucos fall within our demarcation.[42]

Even in the face of the more accurate geographical knowledge presented to the conference by the Portuguese negotiating team, such intellectually authoritative arguments ensured that they were unable to prove conclusively their possession of the Moluccas at Badajoz-Elvas, a situation which left Charles able to reach a financial settlement at Saragossa which, geographically speaking, he should not really have been able to negotiate so successfully.

However, as Argensola's comments testify, Castilian attempts to ensure that the Moluccas were positioned within their demarcated half of the globe did not end with their manipulation of the tenets of classical geography. More contemporary geographical evidence was also gradually amassed to support Castile's claim to the islands. As early as 1522 Nuño Garcia, the Castilian geographer who had drawn several maps for Magellan's expedition,[43] acting on information provided by Elcano, completed a map of the voyage which placed the Moluccas firmly in the Castilian sphere (illus. 42). On this map the line of demarcation is shown running through 'Camatra' (Sumatra), to the west of the islands. Garcia's place at the negotiating table at the initial conference at Badajoz-Elvas suggests that if not this very map, then maps based on its calculations were offered in support of Castile's claim to the Moluccas. However, the most consummate cartographic manuscript which seems to have been used by the Castilians did not emerge from Garcia, but from his colleague and official collaborator at Badajoz-Elvas, Diogo Ribeiro.

Between his involvement in the original planning of Magellan's voyage and the diplomatic dispute over the Moluccas, Ribeiro had been busy compiling a series of world maps which were to have a profound effect on the controversy. By 1527 he had produced the first of his comprehensive manuscript maps of the world, and by 1529 he had completed a second, even more detailed, world map (illus. 43), on the eve of the successful conclusion of the dispute between Portugal and Castile in Saragossa. Both have been regarded by historians of cartography as some of the most beautiful examples of the art of manuscript mapmaking.[44] In line with the outcome of the previous negotiations carried out at Badajoz-Elvas, Ribeiro's map of 1529 reflected the increasingly fine distinctions which were by this stage separating the two crowns. However, as can be seen from a closer scrutiny of its western section, Ribeiro proceeded to place the Moluccas just within the Castilian sphere, a striking example of the extent to which he was aware of how sensitive was his placement of the islands. Ribeiro positioned them 172° $30'$ W of the Tordesillas line of

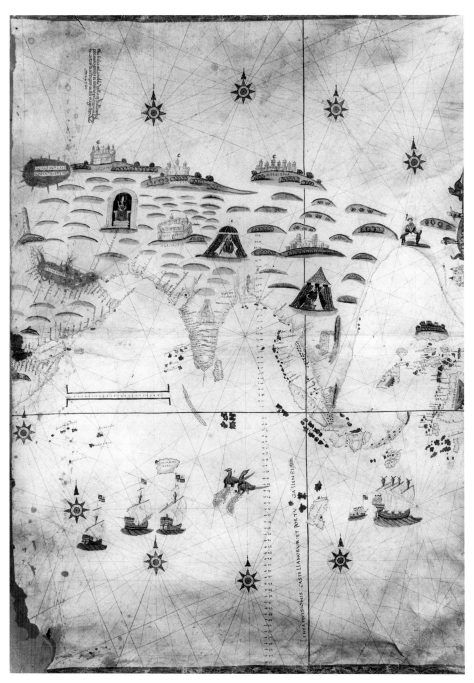

42 Nuño Garcia, chart of the Moluccas, *c.* 1522, vellum. Biblioteca del Re, Turin.

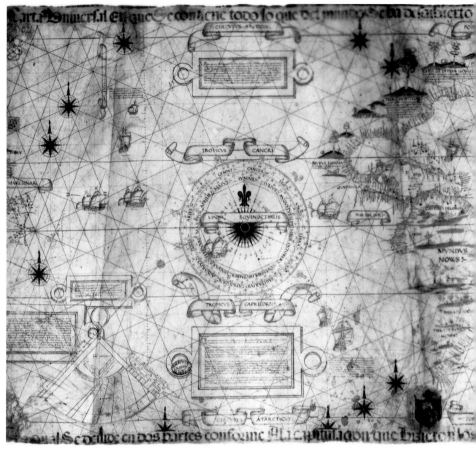

43 Diogo Ribeiro, world map, 1529, vellum. Biblioteca Apostolica Vaticana.

demarcation, just 7½ degrees inside the Castilian sphere.[45] This broke with all previous geographical representations of the islands, which had repeatedly placed them in the eastern (Portuguese) hemisphere. For the very first time Ribeiro's map sensationally placed the Moluccas in the western hemisphere. Overall the map offers an accurate representation of the Atlantic but Asia is slightly overextended, echoing Ptolemy, a convention which Ribeiro would have known from his earlier cartographical studies carried out in Portugal to be wholly inaccurate.[46] Even more significantly, the breadth of the Pacific is greatly distorted on his map, with the distance between the westernmost point of South America and the Moluccas comprising just 134 degrees, which was a substantial underestimation of the actual distance between the two points.[47]

Ribeiro had skilfully managed to produce a map in line with the rapidly changing requirements of the Castilian claim to the Moluccas, and

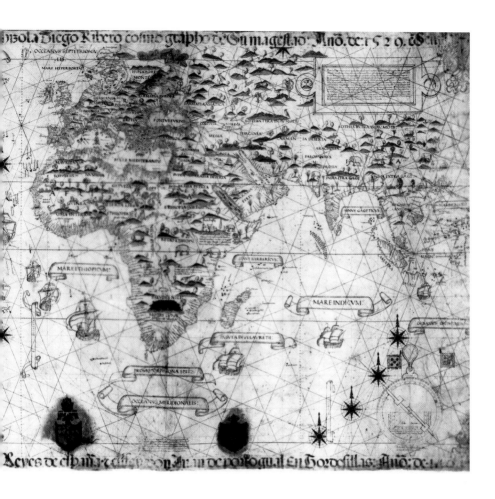

manipulated the enormous amount of highly sophisticated geographical material at his disposal, moving as he had between Lisbon and Seville. What he produced was not only a map which reflected the most up-to-date geographical knowledge then available, but one which also emphatically claimed the righteousness of Castile's claim to the islands which it prized so greatly. Whilst it is not clear as to whether or not Ribeiro's map was presented in evidence at Saragossa in 1529, its monumental composition suggests that it formed the basis for those maps which were presented, and its continuing geographical influence stood as an abiding image of Charles' claim to the Moluccas, which could be quickly consulted if the emperor felt the need to reclaim his rights to the islands.

The dispute between Castile and Portugal provided a unique opportunity for the establishment of what could be called a 'tournament of value'[48] between the geographers and their maps as they jockeyed for diplomatic and geographical supremacy. Figures like Ribeiro and Garcia shrewdly worked within the political confines of the roles assigned to them in the midst of the controversy over the Moluccas. What they achieved was the enhancement of their own social status as professional geographers, whose cultural and political reputations were enhanced by their successful participation in a tournament of value which saw their products exhibited as geographically instructive and politically persuasive cultural artefacts. What was even more significant in Ribeiro's case, was his ability to utilize scientific uncertainty to ensure that his maps could not simply be dismissed as propaganda. The inability to measure longitude objectively ensured that the exact position of the Moluccas could not be settled scientifically for more than 200 years. Ribeiro therefore skilfully exploited scientific uncertainty over the measurement, both to support Castile's claims to the islands and to insulate his own world map from charges of wilful geographical distortion. Ribeiro's map could therefore only be attacked by maps which offered an even more comprehensive mix of political, commercial, intellectual and scientific authority. However, Ribeiro had established his own geographical reputation through exactly this ability to draw on a range of knowledges, particularly practical navigation and the learned geography of classical authors like Ptolemy, in producing a map which managed to satisfy almost everybody. It satisfied seaborne pilots, for its apparent hydrographic 'accuracy'. It satisfied academic cosmographers, for its adherence to the classical geographers. It satisfied the Castilian negotiators, for its support of their claim to the Moluccas, and it even satisfied the Portuguese for its acknowledgement of their own discoveries, in all but one crucial aspect – its positioning of the Moluccas. Operating beyond the routine parameters of geographical enquiry, Ribeiro seized the chance to display his map to a range of disparate social groups which would not have otherwise consulted his labours. His success ultimately lay in his ability to create the impression of scientific objectivity and hydrographic accuracy in a map which was deeply implicated in the Castilian claim to the Moluccas, employing the rhetoric of classical learning in the pursuit of specific political objectives which were themselves ultimately commercial. Like the 'painted map' forwarded to the Austrian archduke Ferdinand by Charles V in 1523, Ribeiro's map consummately participated in the trading of commercially useful territories between competing crowns, inhabiting a space which fused territorial claims with commercial ventures, a process which only served to enhance the status of Ribeiro's own maps as themselves diplomatically and commercially valuable objects.

Ribeiro's state salary of 30,000 *maravedis* must have been a small price to pay for the advantage to Charles of cashing in his claim to the Moluccas in return for 350,000 gold ducats. Professional geographical expertise came cheaply when such amounts were at stake. For the emperor the financial return on such a deal was clearly more attractive than establishing possession of the far-flung islands. The advantage of a quick return on the stake made on them obviously seemed far more attractive to Charles and his creditors than the costly upkeep of administering the islands and the spice trade which emanated from their precious soil. By May 1529 confirmation of his successful negotiation to wrest such a large sum of money from the Portuguese crown had reached the ears of French merchants, for whom the injection of a substantial amount of money into Charles' coffers could only be good for business. On 6 May, Gaspar Contarini wrote to the Venetian seigniory announcing that:

The merchants of Lyons had news from Spain of the conclusion of the bargain between the Emperor and the King of Portugal about the navigation of the Indies, and that thereby the Emperor would obtain from 300,000 to 350,000 crowns.[49]

Viewed within this context, the geographical manipulations of Encisco, Garcia and Ribeiro should not be viewed as failures. Whilst they did not establish Castilian possession over the Moluccas, it would seem that from a very early stage in negotiations this was never Charles' real aim in contesting the sovereignty of the islands. Instead the maps, charts and navigational calculations of the geographical specialists attached to the Castilian negotiating teams were vital to securing his financial security in the pursuit of imperial adventures nearer to home.

The diplomatic and geographical contest between Castile and Portugal over the Moluccas has invariably been disregarded within both histories of cartography and more general historical accounts of the politics of early sixteenth-century Europe. Critical attention has tended to focus on the significance of the 1494 Treaty of Tordesillas in defining the geographical scope of the imperial pretensions of the crowns of early modern Europe. However, the treaty only provisionally defined the outward expansion of the late fifteenth-century voyages. The Treaty of Saragossa was far more significant in literally completing the circle of the increasingly global pretensions of the crowns of Portugal and Castile by finally joining up the eastward and westward voyages of the early modern period to create the definitive global image which can be seen in Holbein's portrayal of the terrestrial globe in *The Ambassadors*. The fact that this global image was finally established in the territories to the east of early modern Europe is symptomatic of the extent to which such eastern horizons have been marginalized in critical accounts of the period, which have focused almost exclusively on the 'discovery' of the New World by Columbus.

However, it was the search for the territories to the east which inspired not only Columbus' first voyage in 1492, which resulted in the Treaty of Tordesillas, but also Magellan's circumnavigation of the globe in 1522, which ultimately resulted in the Treaty of Saragossa. In sailing west to reach the east, voyagers like Columbus and, even more significantly, Magellan exposed the limits of maps and charts which portrayed the earth upon a flat surface and ushered in a new era of global geography.

The impact of this increasingly global geographical vision on mapmaking and the study of geography was momentous, and sparked off a whole new industry of globemaking as merchants and princes demanded access to the world picture that began to come into increasing focus in the aftermath of Magellan's voyage and the negotiations over the Moluccas. In 1530 the renowned geographer Gemma Frisius began working on the creation of terrestrial globes, and in 1535 produced an engraved globe in Louvain,[50] under the direct patronage of Charles V (illus. 44). It is worth invoking Frisius' comments on the significance of these terrestrial globes within the context of the Moluccas dispute, to emphasize the political importance of this new global geography:

The utility, the enjoyment and the pleasure of the mounted globe, which is composed with such skill, are hard to believe if one has not tasted the sweetness of the experience.[51]

Charles clearly 'tasted the sweetness' of Frisius' global geography: like Ribeiro's world map from which it drew much of its information, it placed the Moluccas well within his half of the globe, offering further evidence of Charles' ability to make a renewed claim to the islands whenever he wished.

The dispute over the Moluccas and the subsequent global development of early modern geography ushered in a period of unprecedented geographical awareness in sixteenth-century Europe. This awareness presented geographers and mapmakers with an unprecedented level of social prestige and relative intellectual autonomy. In the late fifteenth century figures like Berlinghieri and Behaim had struggled to establish their regional and global maps as images to which their patrons turned to define their political and commercial place in the early modern world. The increasing sophistication of the maps and globes of geographers like Ribeiro and Frisius allowed the study of geography to go one step further. When issues such as the measurement of a single degree made the difference between financial profit and potential commercial ruin, geographers were given increasing licence to define the parameters of diplomatic disputation over territories and trade routes whose possession was vital to the political authority of the empires of the early modern world. Increasingly employed as advisers and geographical specialists in the courts of

44 Gemma Frisius, terrestrial globe, engraved gores on a plaster base, *c.* 1535, Louvain. Globenmuseum der Österreichisches Nationalbibliothek, Vienna.

sixteenth-century Europe, geographers like Mercator and Ortelius eagerly exploited their new-found freedom to create some of the most significant maps, globes and atlases of early modern geography, and it is this development which concerns the final chapter in this book.

As far as Bartholomé de Argensola was concerned, looking back in 1609 at the diplomatic controversy over the Moluccas:

The most bloody Theatre of continual Tragedies, was *Ternate* and all the *Molucco's*. There both Nations of *Castile* and *Portugal* decided their Quarrel by the Sword, whilst their Kings in Europe only contended by Dint of Cunning, and Cosmography.[52]

The debate over the Moluccas was not only a watershed in the political history of early modern Europe, but it was also a turning point in the political apprehension of the importance of geography in defining the contours of the early modern world. To produce and possess maps and globes, as Charles V knew to his profit and João III to his chagrin, signified not only access to the possibility of laying claim to contested territorial possession, but also imbued their owners with an air of commercial authority. Both of these values were ultimately provided through recourse to the cunning of the cosmographer.

5 Plotting and Projecting: The Geography of Mercator and Ortelius

With the settlement of the dispute over the Moluccas between the Castilian and Portuguese crowns at Saragossa in 1529, a definable shift began to develop in the geographical conception of the early modern world. Magellan's circumnavigation of the globe and the subsequent diplomatic controversy over the Moluccas led to the establishment of a geographical and political distinction between the territories of the so-called 'Old World' to the east of mainland Europe and the territories of the 'New World' to the west. As the Americas came into increasing geographical definition, the concept of the Old World itself began to undergo a definable change. The older traditions based on Ptolemy and medieval T-O maps had established an abiding preoccupation with the territories to the east, a preoccupation which continued to predominantly structure the geography of the early sixteenth century as the Portuguese attempted to bring the commercial markets of the Old World under their jurisdiction. However, this concentration on a directional east still remained relatively free of subsequent cultural and political stereotypes of the dark, mysterious and exotic 'Orient', which came to suffuse the political geography of the Enlightenment and beyond.

However, it took a geographical and diplomatic dispute over the territories of the Old World to begin to circumscribe a directional east in relation to its antithetical west, as the Moluccas dispute brought the Portuguese activities in the Old World right up against the increasingly focused territories of the Americas established by the Castilian advances in the New World. The English diplomat Robert Thorne, who had closely followed the dispute over the islands whilst living in Seville throughout the 1520s, vividly encapsulated this growing distinction between east and west in his report to Henry VIII, written in 1527:

For these coastes and situation of the Ilands every of the Cosmographers and pilots of Portingall and Spayne doe set after their purpose. The Spaniards more towards the Orient, because they should appeare to appertaine to the Emperour [Charles V]: and the Portingalles more toward the Occident, for that they should fall within their jurisdiction.[1]

Lacking a suitable language of directional east and west, Thorne adopted the terms 'Orient' and 'Occident' in an attempt to come to terms with the expansion of the two crowns.

As diplomats like Thorne struggled with the implications of this confusing, and still not as yet fully geographically defined, distinction between Orient and Occident, it became incumbent upon the practising geographers and mapmakers of the period to try to produce a coherent representation of the globe that distinguished east from west, and also to establish where Europe as a geographical and political entity fitted into this changing world picture. This chapter traces the responses to this problem produced by two of the most renowned geographers in the history of early modern cartography: Gerard Mercator and Abraham Ortelius. I want to suggest that, working in the aftermath of the dispute over the Moluccas, it is no accident that both these geographers redefined both the geographical representation of the world, and also the status of geography in the early modern period. Both have been rightly celebrated for their contributions to the establishment of geography as an intellectually and politically respected discipline, establishing ways of creating and presenting geographical information which continue to influence the field of geography even today. Whilst Mercator's famous method of projecting the globe on a plane still remains the most widely used technique of representing the earth's surface, Ortelius' method of compiling and endlessly updating maps and geographical information within the conveniently portable framework of a printed atlas also remains the standard by which contemporary geographers package and present their data. Due to the veneration which has been accorded both men over the centuries for their solutions to the problems of geographical representation and distribution, the social and political determinants which shaped their products have often been overlooked in an attempt to establish their intellectual achievements as being those of disinterested scholars, objectively pursuing the establishment of geography as a scientifically rigorous intellectual discipline. However, this response to Mercator and Ortelius fails to appreciate the extent to which their output was produced within the context of the increasingly commercial and political nature of geography and mapmaking in the aftermath of the Moluccas dispute. It was the specific peculiarities of this political dispute, and the geographical controversy which it produced, which to a great extent established the conditions under which Mercator and Ortelius created their most significant maps, globes and atlases from the 1530s to the end of the sixteenth century. This is not to suggest that their output completely sacrificed geographical accuracy in the face of political pressure. Instead, what is most revealing in the work of both geographers is the ways in which they consistently produced politically sensitive but geographically comprehensive maps and globes. Their products managed to satisfy both their political patrons and the demands of the market-place, whilst also magnifying the social status of first Mercator and then Ortelius as revered but also worldly geographical specialists, who were capable of unifying a tangle of disparate and often contradictory geographical observations, reports and travel accounts into aesthetically magnificent and intellectually comprehensive maps and atlases, which moved towards establishing a standard representation of the early modern world.

Printing and Profiting: The Marketing of Printed Geography

One of the most significant aspects of the geography of both Mercator and Ortelius was the extent to which they skilfully adapted the growing medium of print to market their work. The diffusion of printing towards the end of the fifteenth century had already established Ptolemy's *Geographia* as one of the most influential geographical texts available to the princes, diplomats, scholars and merchants of the period. As has already been seen, geographers and editors like Berlinghieri were quick to appreciate the ability of the printing press to accurately and rapidly disseminate a standard representation of the world. However, by the 1520s the improved techniques of printing allowed for an even greater diversification of printed geographical texts, which could be distributed on an even greater scale. Commercial centres throughout northern Europe such as Antwerp, Brussels, Strasbourg, Nuremberg and Augsburg provided the financial and material contexts for presses which located themselves at the crossroads of learning and commerce, the two most important ingredients in the world of printing. The result was an explosion of material on issues as disparate as religion, education, science, travel, philosophy, literature and, of course, geography. Although the impact of print was invariably limited to literate communities who were able to afford printed texts, the sheer number of both written and visual books which came off the presses of northern Europe in the early decades of the sixteenth century indicates the extent to which the influence of print culture extended well beyond the rich and the learned. The printing of maps and books on geography was symptomatic of this dispersal of print throughout sixteenth-century European culture.

The maps and geographical books which came off the presses of Strasbourg at the beginning of the sixteenth century provide just one particularly well-documented example of the spectacular growth in printed mapmaking during this period. The city was home to some of the most innovative editors and printers working in the field of geography.

Between 1500 and 1520 its presses produced no less than seventeen works specifically concerned with geography.[2] In 1507 one of the most influential printers in the city, Johannes Grüninger, published an enormous woodcut, on twelve sheets, of Martin Waldseemüller's famous *Universalis cosmographia secundum Ptholomaei traditionem et Americi Vespucii aliorumque lustrationes* (*A Map of the World According to the Tradition of Ptolemy and the Voyages of Americus Vespucci and Others*; illus. 27). This was the first map to unite the disparate collection of islands which had been thought to constitute the New World into a single landmass, named 'America'. The production of such maps, uniting printed excellence with up-to-the-minute information on the new discoveries, established Strasbourg as a major centre of map printing. However, what was significant about the technology which produced maps of the standard of Waldseemüller's was not just the precision of the maps, but their sheer number. Grüninger printed 1,000 copies of Waldseemüller's wall map which, considering that it was composed of twelve separate sheets, meant a print run of 12,000 separate sheets. He went on to print a series of maps designed by Waldseemüller in 1522 and 1525. Evidence suggests that by 1525 Grüninger had printed no fewer than 3,500 wall maps composed on separate sheets, involving over 98,000 large folio sheets, a staggering investment of time, skill and resources in the relatively novel art of map printing.[3] As a shrewd businessman with contacts throughout the commercial world of sixteenth-century Strasbourg, Grüninger would not have made such an investment in Waldseemüller's work without a firm conviction that a public existed to purchase such beautiful and expensively produced maps.

The ability of printing like Grüninger's to exactly reproduce comprehensive maps such as Waldseemüller's and distribute them on a mass scale to the book (and map) buying public not only expanded a general awareness of the cultural significance of maps, but also led to significant changes in the conditions under which they came to be produced. The part played by printing in reproducing the detailed manuscript maps drawn up by geographers like Berlinghieri and Ribeiro established a perceptible distance between the geographer and his patron. Whilst Waldseemüller acted as the official geographer to the court of René II, Duke of Lorraine, his involvement with Grüninger's press in Strasbourg allowed him a degree of relative intellectual autonomy in the supervision of his work that was not often accorded to scribal scholars and geographers, who invariably undertook their commissions under the very roof of their particular patron. What increasingly defined the intellectual production of scholars and geographers like Waldseemüller was not the contacts established within a closely knit group operating under the patronage of a single figure, but the so-called 'republic of letters', the diffuse collection of scholars, printers and travellers scattered across the states of early modern Europe. The exchange of information and knowledge within this informal republic allowed for a growing diversification of geographical representation, which was aimed not only at a specific patron, but also at an increasingly anonymous book-buying public. As a result, patrons began to associate themselves with particularly respected geographers, rather than vice versa, as had been the case with Berlinghieri, who had repeatedly attempted to associate his 1482 *Geographia* with a politically influential patron. It was under these changing conditions of production and circulation that both Mercator and Ortelius fashioned their own highly distinctive maps, globes and atlases.

The early career of Gerard Mercator (illus. 45) is symptomatic of the changes which were taking place within the field of printed geography by the 1530s, and is worth considering briefly to examine the ways in which he mediated between the political requirements of his cartographic products and the need to market his work skilfully within the context of an increasingly discerning map-buying public. After studying at the University of Louvain, Mercator, like many sixteenth-century geographers, applied himself to the study of the specific technical skills required to produce printed maps and globes. By 1535 he had established himself as a copperengraver of repute, and was responsible for engraving the gores for Gemma Frisius' terrestrial globe, which was produced in the same year. Yet Frisius' globe, as has been seen, was itself caught up in the continuing controversy over the placement of the Moluccas. Having positioned the islands within the Castilian half of the globe, it laid claim to comprehensive knowledge of the terrestrial globe whilst also aligning itself politically with the interests of its patron, none other than Charles V himself. In many ways Frisius' globe was an even more enduring geographical symbol of Charles' claim to the Moluccas than Ribeiro's world map of 1529 as, unlike Ribeiro's map, this globe, made from printed gores engraved by Mercator, could be repeatedly reproduced to underline Charles' claim to the spice islands, a claim which was further strengthened through its representation within the powerful medium of print. In February 1536 Charles issued a privilege, which gave Frisius and the globe's printer, Gaspard van der Hayden, a monopoly on reproducing the printed image of the globe. The privilege is worth quoting in full as an example of the ways in which printers and geographers came to exercise increasing commercial and representational control over their politically sensitive geographical products:

Inasmuch as our loyal Subjects, the Mathematician Gemma Phrisius and Gaspar a Myrica [Gaspard van der Hayden], do purpose to issue for the general

use of those interested, a globe or sphere, enlarged, augmented and more splendid than that previously issued (together with a Celestial Globe), with the addition of countries and islands recently discovered;

And inasmuch as We are favourably disposed to this praiseworthy scheme of theirs by which they endeavour, not only to elucidate Mathematics, revive the memory of ancient kingdoms and conditions, and present them to the people of our age but also to hand down to future centuries the memory of our own age and our own Kingdom, which (by the grace of God) encompasses many islands and territories practically unknown to any previous century, and not a few of which are fortunately being imbued with the Christian religion;

In full knowledge of the facts, We now, by virtue of the document, prohibit and forbid all printers, and booksellers, and all heads, officials or employees of the book-trade, as well as all persons, subjects of ours and of the Holy Roman Empire, whatever the name they bear, and whatever rank, privilege or position they may hold, to duplicate such globes or spheres, or the Introductions and Forewards explaining them, under whatever title they be issued, or may be issued, by the said Gemma and Gaspar jointly or separately, within the next four years following upon the drawing–up of the present document, or otherwise to cause to have made, or cut in metal or other material, or to introduce, sale, or hawk, publicly or privately, globes or introductions made or cut elsewhere, in similar or other form, without the consent of the said Phrisius and Gaspar;

Upon pain of Our severe displeasure, and a fine of 10 marks in pure silver, to be devoted to Our fiscal requirements, besides the forfeiture of the work or books cut, or made and which the said Gemma and Myrica, or their agents, may be entitled to confiscate and use, wherever they be found, with the aid of the authorities or at their own discretion, for whatever purposes they may determine.[4]

What is so significant about the privilege is the extent to which it protects the rights of the producers of the globe on the open market, rather than restricting the reproduction of the globe according to the wishes of its patron, Charles V. Any unauthorized reproduction, once sold on the open market to anyone with sufficient means to purchase it, would obviously minimize any financial profit which Frisius and van der Hayden would reap from sales of their unique globe. Charles was undoubtedly 'favourably disposed to this praiseworthy scheme'; not only did the globe reassert his claim to the Moluccas, but it also flew the emperor's flag over his recent conquest of Tunis in North Africa, which Charles had taken from the Ottomans in 1535. Yet the privilege which Frisius and van der Hayden negotiated in return for such a geographically flattering portrait of the scope of Charles' territorial possessions gave the geographer and printer sole rights to the reproduction of their globe, in an attempt not only to signify its geographical importance as a masterful example of globemaking, but also to maximize its commercial profitability in an increasingly competitive open market.

Evidence of the strategies which Frisius and van der Hayden adopted in the marketing of their terrestrial globe is vividly encapsulated in one of the legends which adorns its surface:

Gemma Frisius, doctor and mathematician, described this work from various observations made by geographers and gave it this form; Gerard Mercator of Rupelmonde engraved it with Gaspard van der Heyden, from whom the work, a product of extraordinary cost and no less effort, may be purchased.[5]

The legend, presumably composed by Frisius, stresses the financial value of the globe in terms of both the intellectual labour which went into its production and the material cost involved in putting together such a complex object. As the master copper-engraver whose plates went to van der Hayden's printing press after their approval by Frisius, Mercator appears to have been fully aware of the ways in which both his teacher Frisius and his printer van der Hayden maximized the commercial and intellectual significance of their globe whilst also creating a geographical artefact which gave considerable pleasure to its patron.

Mercator's ensuing geographical career suggests that he quickly learned from the experience of working with Frisius and van der Hayden. His cartographic output consistently united political caution and technical brilliance with an eye to the expanding commercial market in maps and globes. In keeping with the system of patronage which had served him so well whilst working with Frisius, Mercator went on to engrave a map of Flanders in 1540, which he dedicated to Charles V.[6] His subsequent work brought him in touch with some of the most influential decision-makers in Charles' empire as he created several mathematical and cartographical instruments which he dedicated to Nicholas Perrenot de Granvelle, a member of Charles' privy council and one of the emperor's closest advisers.[7] By the early 1540s Mercator was astutely using his political contacts and technical expertise to fashion an identity for himself as one of the most respected geographers of the day. In 1541, in direct imitation of his former teacher Frisius, he issued his own printed terrestrial globe, the first to appear under his name alone (illus. 46). In the words of his self-appointed biographer Walter Ghim, Mayor of Duisburg, writing in 1595, Mercator:

Dedicated it [the globe] to a distinguished nobleman, M. Nicholas Perrenot de Granvelle, by far the most eminent of the Privy Council of the Emperor Charles V. Meanwhile, through the recommendation of this gentleman, he came to the notice of the said Emperor, and he made for his Majesty a large number of scientific instruments of exquisite workmanship.[8]

Like Frisius, Mercator took care to ensure that his globe obtained an imperial privilege from Charles forbidding its unauthorized reproduction, along similar lines to the one granted to Frisius just five years earlier.

In his description of the construction of this, his first globe printed and marketed under his own name, Mercator carefully weighed up not only the geographical rationale for producing it, but also the financial reasons which underpinned his decision to undertake the creation of such a time-consuming artefact. In a letter to Granvelle's son Antoine, he explained his reasons for making the globe:

As I had anticipated before I began work, I have concentrated my attention on the 'spherica geographia' so as to have a shield for my domestic financial obligations. Consequently, I began to devote spare hours to comparing the old geography with the new. The more carefully I examine, the more errors I find in which we are enmeshed. It seems particularly erroneous to take Malacca for the Aurea Chersonesus (the 'Golden Peninsula') of Antiquity; moreover, they make Taprobana into an area very close to Malacca and only separated from it by a relatively narrow strait; Ptolemy placed it about 30 degrees from the great circle away from China. And yet, our hydrographers have placed Taprobana not further away from Cape Similla, now Cumari [Cape Comorin] than Ptolemy did. They consider the length of India from west to east to be half that of what Ptolemy suggested, and yet, Strabo accepted Ptolemy's length, based on solid grounds. When we, blinded in this way, attempted to harmonise the irresolvable difference between the old and the new, we denounced both the ancient and the more recent descriptions; in addition, by means of small adjustments, we undermined the current symmetry of the coasts as well as the findings the ancient geographers had achieved through great effort. How deceived we are in our

46 Gerard Mercator, terrestrial globe, engraved gores on a plaster base, 1541. Kultur- und Stadthistorisches Museum, Duisburg.

own conception of the Far East will be sufficiently clear to anyone who attentively reads Marco Polo the Venetian. For this reason, in the hope that our conclusions of this type will fall upon fertile ground with those who are interested, I have decided to publish a terrestrial globe. Secretary Morillon and Adrianus Amerocius are extremely pleased with this decision and encourage its being carried out.[9]

Mercator's comments are a fascinating insight into the commercial and geographical considerations which preoccupied sixteenth-century geographers working in the medium of print. It appears to have been more financially lucrative for him to produce the type of globe which, in the aftermath of the Moluccas dispute, was in great demand. Before embarking on any grand geographical project, Mercator had to ensure that his labours would provide him with a large enough financial reward to 'shield' him from 'domestic financial obligations'. Significantly, he noted that the project had already received the political support of the emperor's political advisers, Granvelle and Morillon both being 'extremely pleased' with his globe. For Mercator, it was clearly more commercially viable to produce a globe which had received the sanction of the crown than to embark on a detailed series of printed maps whose commercial success could in no way be guaranteed. Yet even more tellingly, in his discussion of the geographical innovations which his globe would offer, he covered exactly the same ground as that pursued by the pilots and geographers involved in the Moluccas dispute fifteen years earlier. His apparently respectful attempt to compare 'the old geography with the new', which involved a comparative analysis of Ptolemy's geographical accounts with 'our hydrographers', was in fact a judicious intervention within the terms of the controversy over the Moluccas and their position in relation to the older Ptolemaic calculations. The main points of disputation surrounded the erroneous positioning of the Portuguese-held Malacca, the placement of Taprobana (modern Sri Lanka) too far to the east, and the relative 'length of India from west to east'. All of these disputes were, as has been pointed out, central to the political dispute over the Moluccas.

Mercator's commercially and politically sensitive response to the controversy over the size and extent of the Ptolemaic Old World also emphasizes the way in which he carefully fashioned his own intellectual identity as an impartial scholar engaging in an academic inquiry 'to harmonise the irresolvable difference between the old and the new' geographical positions, which had been so clearly exposed in the dispute over the Moluccas. Steeping himself in the writings of Strabo and Ptolemy, Mercator was nevertheless astute enough to combine such learning with carefully judged interventions into more contemporary political issues, without appearing to take sides. The new market in printed maps and globes not only desired a geography imbued with the classical tradition of Ptolemy and Strabo, but it also required a level of ostensible impartiality in the geographical disagreements which had characterized the diplomatic disputes between Castile and Portugal regarding overseas territories since the Treaty of Tordesillas in 1494. The manuscript maps of Ribeiro and Garcia had been produced for a specific political moment in time, and despite their success were not destined for acceptance into the pantheon of a geographical discipline which was increasingly wrapping itself in the mantle of a purportedly scientific objectivity. Mercator's ability to fashion an intellectual identity characterized by the pursuit of supposedly objective learning ensured that he retained the aura of a disinterested geographical savant, whilst simultaneously ensuring the retention of a level of imperial patronage and commercial success which was to ultimately secure his future economic independence.

Mercator's ability to combine geographical skill with an astute management of the commercial and political implications of his work quickly began to reap intellectual and financial rewards, as his products became some of the most sought after in sixteenth-century Europe. In 1568 the Spanish diplomat Benito Arias Montano wrote to his superiors in Spain, informing them that the globes of both Frisius and Mercator could be purchased from the printing houses of Antwerp. The Frisius globe of 1535 could be purchased for eight escudos, whilst Mercator's 1541 globe could be bought for twelve escudos. Montano's superiors subsequently ordered him to buy the latter, an indication of the reputation that Mercator's globes were beginning to attract by the late 1560s.[10] From the 1550s Mercator's sales were managed by one of Antwerp's most respected printers, Christopher Plantin. Between 1558 and 1576 Plantin's business ledgers record that he sold no fewer than 868 copies of Mercator's wall map of Europe, which had been printed in 1554.[11] Again, in creating this particular map Mercator took care to dedicate it to his imperial patron, Antoine Perrenot.[12] His growing reputation had earlier brought him into contact with Charles V himself, and in 1552 Mercator had:

Constructed by order of the Emperor two small globes, one of purest crystal and one of wood. On the former, the planets and the more important constellations were engraved with a diamond and inlaid with shining gold; the latter, which was no bigger than the little ball with which boys play in a circle, depicted the world, in so far as its small size permitted, in exact detail. These, with other scientific instruments, he presented to the Emperor at Brussels.[13]

This apparently rather whimsical commission only further strengthened the status of geography as a discipline which provided emperors and princes with tangible evidence of their access to an understanding of not only the earth, but also the stars. Mercator elegantly expressed his ability to reveal the arcane mysteries of the cosmos in the production of the celestial globe, whilst also emphasizing his ability to operate at the more worldly and political level of the terrestrial globe. The crystal, gold and diamonds which made up Mercator's celestial globe were only further evidence of the riches which could be extracted from the distant places represented on the terrestrial globe, territories which had been made accessible by geographers like Ribeiro and Garcia. Mercator appears to have been happy to cash in on this situation as a way of increasing his esteem at Charles' court, whilst also proceeding to work on even more ambitious geographical projects.

Politics and Projections: Mercator's 'Map of the World'

By the early 1560s commissions such as those undertaken for Charles V allowed Mercator the financial security and relative intellectual freedom to embark upon an even more grandiose project than the terrestrial globe which he had produced in 1541. The problem which he undertook to resolve was one which had troubled sailors for decades. Navigators had always been aware of the unfeasibility of using terrestrial globes to provide suitable explanations of how to navigate simply and effectively across open seas on long-distance voyages. Whilst globes such as those produced by Frisius and Mercator struggled over the geographical implications of uniting the Occident with the Orient in the aftermath of Magellan's voyage, their size and bulk prevented their effective utilization at sea. Although they do appear to have been part of the esoteric geographical paraphernalia taken on overseas expeditions, plotting a course across a spherical globe was simply not viable. Sailors required rutters (written sailing instructions) and, increasingly, flat charts across which they could accurately and confidently project their oceanic progress day by day.

However, as the Portuguese had known for decades, flat portolan charts of the type used by sailors in the Mediterranean were wildly inaccurate when faced with long-distance voyages as they failed to allow for the natural curvature of the earth. Tracing a line between two points on the earth's surface involved taking this curve into account. Lines of bearing on portolan charts were unable to solve this mathematical conundrum, and as a result traditional sailing along an apparently straight line between A and B would consistently lead pilots slowly off-course, as they failed to allow for the fact that the line of bearing curved gradually due to the spherical nature of the earth. The further the distance travelled, and the longer the line of bearing pursued, the more ships sailed further off-course. Geographers therefore went in search of a geographical projection which would literally square the circle by portraying the sphericity of the globe on a flat surface. This would in theory allow pilots to plot their courses across the surface of a flat map along a straight line of constant bearing which, as the map took into account the curvature of the earth, would enable them to reach their appointed destination without sailing off-course. By the middle of the sixteenth century such a method of navigating was becoming increasingly necessary to keep pace with the growing intensity and nature of long-distance travel. Sailing into the relative unknown and discovering previously uncharted land no longer motivated long-distance seaborne travel. The commercial ethos which surrounded both the Portuguese and Castilian ventures into the waters of the Indian and Pacific Oceans ensured that it was necessary to be able to navigate rapidly and accurately between a specified point of departure and a specified point of arrival. It became increasingly imperative for empires and merchant-financiers investing in long-distance enterprises to ensure that expensively outfitted expeditions reached their appointed destinations quickly and returned as soon as possible to maximize the financial profits from their precious cargoes. Bound up with such commercial concerns were political considerations which increasingly restricted the arbitrary movement of ships across the open seas. The Treaty of Saragossa had nominally circumscribed the geographical limits within which shipping could move. The diplomatic repercussions of any significant disclosure of maritime transgressions of these lines of demarcation, however innocent, were nevertheless potentially explosive.

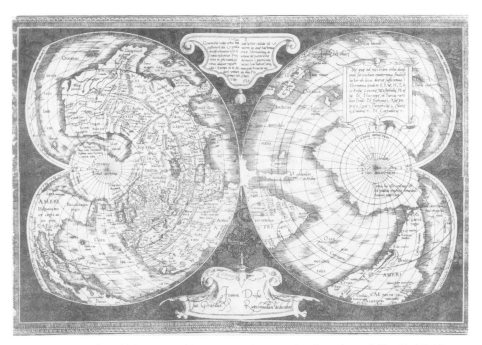

47 Gerard Mercator, world map, copperplate engraving, Louvain, 1538. New York Public
Library.

Mercator had already experimented with the problem of mapping the
globe on a plane surface in 1538, when he published a world map on a
heart-shaped projection (illus. 47). However, this was quite clearly of little
use to navigators and pilots. It was in response to this problem that he
began work on a new world map on a plane surface which would account
for the curvature of the globe, and which would allow pilots to chart a
straight line of navigational bearing, free from the distortions of the older
portolan charts. After many years working on this new map and the complex mathematical projection required to emplot the spherical coordinates
of the earth's surface onto a flat map, Mercator published his now famous
Nova et aucta orbis terrae descriptio ad usum navigatium emendate accomodata
(illus. 48). The intellectual significance of this new world map was underlined by the care which Mercator again took in the marketing of the new
projection. He was granted two royal privileges protecting his rights over
reproduction of the map, one from Charles V to cover its publication and
circulation in the German states for fourteen years, and one from Charles'
son Philip II which forbade its unauthorized reproduction for ten years
throughout the Netherlands.[14] The whole map was dedicated to Mercator's patron, 'the most illustrious and clement Prince and Lord William,
Duke of Julich, of Cleves and of Berg'.[15] The care which Mercator took

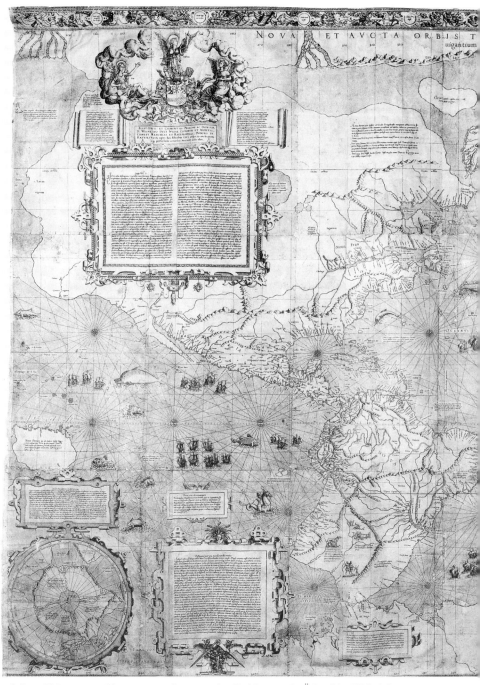

48 Gerard Mercator, world map, copperplate engraving, Duisburg, 1569. Öffentliche Bibliothek der Universität Basel.

over the rights to its production and distribution was indicative of the importance he attached to a map which was to revolutionize not only sixteenth-century geography, but ultimately the future development of modern western cartography.

In his introductory notes to the map, Mercator outlined its ambitious nature, reiterating its importance in finally encompassing the global complexity of the earth's surface into a single map represented on a plane surface. He claimed that his intention was:

> To spread on a plane the surface of the sphere in such a way that the positions of places shall correspond on all sides with each other both in so far as true direction and distance are concerned and as concerns correct longitudes and latitudes; then, that the forms be retained, so far as is possible, such as they appear on the sphere.[16]

The difficulty of Mercator's project lay in constructing a plane representation which depicted the meridians as parallel to each other rather than, as is the case with the true representation of the globe, converging on the north and south poles. If this could be achieved, then it would be possible to chart across its surface a line of constant bearing that was straight, rather than a spiral as would be the case when trying to trace it on a globe. The importance of Mercator's innovation in terms of accurate navigational practice and commercial profit was quite clear. Instead of taking awkward and imprecise bearings on board ship across the surface of a globe or a portolan chart, his new projection allowed for a line of bearing to be drawn accurately across the surface of a plane map, explicitly foregrounding, as its title suggests, its usefulness to the art of navigation.

With pilots and navigators in mind, Mercator went on to outline the mathematical procedure which allowed him to emplot an accurate grid of straight lines across his map, whilst also retaining the relative geographical accuracy of the topography of the globe. In his notes to the map, Mercator explained that this new projection involved having:

> Progressively increased the degrees of latitude towards each pole in proportion to the lengthening of the parallels with reference to the equator; thanks to this device we have obtained that, however two, three or even more than three, places be inserted, provided that of these four quantities: difference of longitude, difference of latitude, distance and direction, any two be observed for each place associated with another, all will be correct in the association of any one place with any other place whatsoever and no trace will anywhere be found of any of those errors which must necessarily be encountered on the ordinary charts of shipmasters, errors of all sorts.[17]

This lengthening of the degrees of latitude enabled the globe to be flattened out whilst still allowing for accurate straight lines of navigational bearing to be drawn across the surface of the new map. However, in devel-

oping such a groundbreaking geographical projection, Mercator's calculations were also clearly informed by political considerations as to which
areas were to be represented with the highest degree of topographical accuracy. Whilst the projection allowed for detailed and highly accurate
depictions of the territories radiating outwards from the equator, those in
higher latitudes became increasingly distorted, to the point at which, as
can be seen on the map, the northern and southern polar lands could no
longer be represented with any semblance of geographical accuracy, as
they stretch into infinity. However, this distortion of a north–south axis,
at the expense of a more comprehensive depiction of an east–west axis,
was clearly not one which unduly affected the diplomatic and commercial
preoccupations of Charles V's empire by the late 1550s. By this time the
Habsburg empire had been split into its respective German and Spanish
dominions, and the reach of the Habsburgs stretched right round the
globe, from the Americas to the Philippines. The establishment of such
an extensive global commercial network in the aftermath of the dispute
over the Moluccas came to be reflected in not only the geographical representation but also the mathematical calculations which went to make up
Mercator's new map. This network stretched from Portuguese control of
merchandise flowing back into northern Europe from Southeast Asia and
the Indian Ocean in the east, to the Habsburg involvement in the Americas
and the Pacific Ocean to the west. Mercator's map emphasizes that this
east–west axis was to be the abiding focus of the sixteenth-century world
picture, to the exclusion of a north–south vision of the world which no
longer held any territorial or commercial interest for either the growing
Habsburg or the waning Portuguese empires.

Mercator conceded that his map was primarily concerned with the eastern and western dominions of the Portuguese and Spanish empires, as he
notes in his commentary:

The second object at which we aimed was to represent the positions and the
dimensions of the lands, as well as the distances of places, as much in conformity
with very truth as it is possible so to do. To this we have given the greatest care,
first by comparing the charts of the Castilians and of the Portuguese with each
other, then by comparing them with the greater number of records of voyages
both printed and in manuscript. It is from an equitable conciliation of all these
documents that the dimensions and situations of the land are given here as
exactly as possible.[18]

The 'equitable conciliation' which Mercator's map established emerged
from direct consultation of Castilian and Portuguese maps, which had
been predominantly concerned in previous years with establishing the
position of the Moluccas Islands. Whilst this new map avoided taking
any explicit political position in relation to this debate, it nevertheless

reproduced the increasingly bifurcated distinction between the Spanish-dominated west and the Portuguese-dominated east which began to materialize with increasing political and geographical clarity in the aftermath of the Moluccas dispute. Mercator's projection has been accused of constituting a defining act of Eurocentrism, in its portrayal of an overweeningly prominent Europe placed in the midst of a world picture which scales down the extent of the landmasses of America, Africa and Asia. Whilst the nature of the projection did indeed lead to a relative reduction in the size of these continents in relation to Europe, its conceptual orientation is arguably more complex than simply instating the centrality of Europe. More compellingly, Mercator's world map established a distinction between a geopolitical east and west which reflected their growing polarization in line with the territorial and commercial interests of the sixteenth-century imperial powers. Rather than working outwards from the geographically and politically privileged space of Europe, it oriented its appearance around the continents to the east and west which were claimed to be under the putative control of the competing European empires. It was no longer necessary for a geographer like Mercator to make politically expedient decisions in his geographical calculations in order to satisfy the Habsburg authorities who protected the commercial rights to his maps. As the Spanish empire continued to concentrate its commercial interests in the Americas to the west of Europe, these authorities were presumably quite happy to appreciate a revolutionary world map which defined their sphere of interest in distinct geographical opposition to those of the Portuguese in the east.

The gradual waning of Portuguese authority in Southeast Asia towards the end of the sixteenth century led to the increasing political and commercial predominance of the so-called 'Atlantic World' in European affairs, which itself led to the subsequent political marginalization of the territories to the east of Europe. It was this bifurcation between east and west which I would argue created the conditions for the discursive deployment of the idea of the 'Orient' within European travel accounts and geographical discourse of the seventeenth and eighteenth centuries, which implicitly framed descriptions of an exotic, indolent and mysterious 'East' in relation to a dynamic and enlightened 'West'.[19] These accounts increasingly took as their ontological and geographical centre not the Mediterranean, but the Atlantic. This shift in the geographical and political imagination of the early modern world became enshrined in Mercator's map, which gradually became the model from which all subsequent world maps took their lead. Although the enormity of its scale ensured that it was in fact useless for navigation, the principles of Mercator's projection were utilized in regional maps and charts drawn on a smaller scale to allow for accurate navigation over large distances. Due to this fact alone, the shape of the world defined by Mercator remained the standard which all others imitated.

Rather than establishing the beginning of early modern cartography, in many ways Mercator's map signalled its end. Mercator had produced a world map which would appear recognizably 'modern' to any student of geography today. It had emerged from the diplomatic and territorial disputes between the Castilian and Portuguese crowns which stretched back as far as the Treaty of Tordesillas in 1494, and it ultimately participated in the political and commercial triumphs of the westward-looking Habsburg empire as it consolidated its power base in Spain under Charles' son, Philip II. Mercator's map presented the Spanish-dominated territories in the west on an equal footing with the Portuguese ones to the east which had been the focal point of the cultural, commercial and diplomatic encounters of states such as Venice, Portugal and the Ottoman dominions for centuries. In redefining the geographical and political concept of the 'West' as a commercially powerful, territorially meaningful place, it ushered in not only the predominance of the Atlantic World, but consequently the marginalization of what has come to be defined as the 'East', or the 'Orient'. As the Spanish-based empire of Philip increasingly asserted its ascendancy over the waning power of the Portuguese, it was this Atlantic World, so central to Mercator's map, which was to claim economic, political and cultural supremacy over its geographical and cultural antithesis, the east. Rather than being seen as a product of the monolithic institution of Eurocentrism, Mercator's map should therefore be examined as a carefully negotiated geographical artefact, which created the conditions for the subsequent derogation of the territories of the Old World that had been so central to the commercial, diplomatic and geographical transactions which had characterized the shape of the early modern world since at least the early fifteenth century. Emerging from the diplomatic and geographical disputations which had arisen concerning the territories to the east of Europe, it could be viewed as the first map to signify, geographically and politically, the 'triumph of the west', partly as a result of its comprehensive ability to both define and distinguish this concept of the west from its increasingly demonized antithesis: the east.

A Library Without Walls: Ortelius' Theatrum Orbis Terrarum

If Mercator was responsible for mobilizing an image of the world profoundly affected by the growing global aspirations of the House of Habsburg, then it was his equally famous scholarly contemporary Abraham Ortelius who spent much of his geographical and commercial career marketing his own version of this world picture. Mercator may have been the self-styled Ptolemy of sixteenth-century geography, but it was Ortelius who set out self-consciously to portray himself as its 'capable shop-keeper'.[20]

Like Mercator, his initial involvement in geography was in the commercial production and distribution of cartographical artefacts, rather than in the rarefied field of scholarly geographical enquiry. In 1558 Ortelius, a native of Antwerp, is recorded as having purchased maps from Mercator's publisher Plantin, which he subsequently coloured and resold at a profit. This was a commercial aspect of the map trade which Ortelius came to dominate in subsequent years, and from where he eventually established enough influence to launch a career as a cartographer in his own right. It seems fitting that, considering their mutual business interests in the printing industry, Mercator and Ortelius should have met for the first time at the Frankfurt Book Fair in 1554, and subsequently travelled together on business throughout France in 1560.[21] It is not clear how the more prosperous Mercator financed his trip, but it is significant that Ortelius, who was ostensibly selling copies of the hand-painted maps in which he specialized, financed his trip by using his status as a geographical specialist to conduct commercial business on behalf of his friend Gilles Hooftman, one of the most successful merchants in late sixteenth-century Antwerp.[22]

Ortelius' subsequent career was characterized by his commercially astute ability to commission the engraving and printing of a series of regional maps of Europe for sale on the open market. Map-buyers increasingly demanded that such maps be ever more detailed, reflecting the growing complexity of the changing political geography of late sixteenth-century Europe. Engraved by skilled craftsmen and printed by Plantin, these maps went under Ortelius' name on the basis of his ability to calculate a gap in the market and collate maps which were able to meet the specific demands of an increasingly geographically literate community of scholars, merchants and diplomats. Throughout the 1560s, as Mercator worked on his world map, Ortelius began to develop a scheme to unite the rapidly growing body of regional and global geographical knowledge at his disposal into one volume, an *atlas* of contemporary geography. Whilst earlier printers and editors had redrawn Ptolemy's maps and printed them in a single volume, they made little explicit attempt to market such a product as being of prime importance in interpreting and defining the contemporary world of political geography. However, Ortelius was keen to stress that his new geographical product did just that; it was to be an indispensable tool in understanding the social and political changes which were transforming the society of late sixteenth-century Europe. His colleague Johannes Raedemaeker emphasized the political and commercial considerations which underpinned the production of this new geographical product:

[Ortelius] also bought all the geographical maps that could be had for the sake not only of calculating from the distances, the freight of merchandise and the dangers which they were exposed to, but to estimate the daily reports regarding the European Wars.[23]

Raedemaecker also explained that Ortelius' decision to construct a new atlas based on these principles was primarily inspired by the commercial requirements of one of the geographer's most influential financial backers, Gilles Hooftman. Hooftman had complained to Ortelius that the use of single-sheet maps for business had become a costly and cumbersome process, as large single-sheet maps were difficult to consult quickly and lacked geographical standardization. What merchants like Hooftman required was a compact, comprehensive, up-to-date collection of maps which could range from the regional to the global in providing specific information crucial to the rapid and effective movement of goods and merchandise.

It was on these terms that, in May 1570, Ortelius finally published his monumental atlas under the Latin title *Theatrum Orbis Terrarum*: 'A Theatre of the Terrestrial World'. In its scope and execution it was even more ambitious than Mercator's world map, which had been published the previous year. Along with an address to the reader written by Ortelius extolling the virtues of his *Theatrum*, the volume contained a *Catalogus Auctorum* which provided a list of the eighty-seven cosmographers, geographers and cartographers associated with the project. This was followed by the *Typus Orbis Terrarum* (illus. 49), an updated world map, expunged of the more glaring mistakes enshrined in Ptolemy's *Geographia*, which shared affinities in its geographical orientation with Mercator's map. There followed maps of America, Asia, Africa and Europe, then fifty-six maps of the regions of Europe, six of those of Asia and three regional maps of Africa. Containing a total of seventy maps on fifty-three leaves, the lavishly illustrated *Theatrum* also contained detailed supplementary geographical accounts of the regions represented and their history. Ortelius' *Theatrum* was without doubt the most comprehensive and up-to-date geographical text to emerge within the sixteenth century, a triumph of both the printing press which produced it and the scholarly community throughout Europe which, it was claimed, participated in its production.

Commercial success and scholarly praise for the *Theatrum* were immediate. The initial print run sold so well that a second edition was printed almost immediately. Copies sold for Fl. 6 10st., although with an astute eye for the market Ortelius also provided illuminated printed copies which sold for Fl. 16. To give some idea of the relative cost of the atlas, by

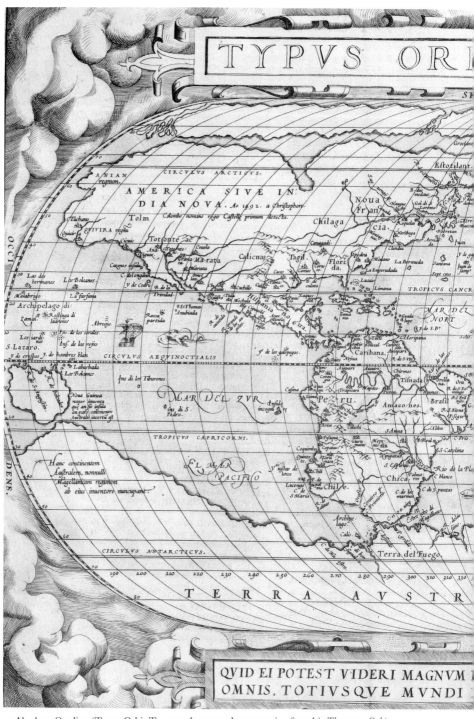

49 Abraham Ortelius, 'Typus Orbis Terrarum', copperplate engraving from his *Theatrum Orbis Terrarum*, Antwerp, 1570. British Library, London.

1580 an unbound, unilluminated copy could be purchased for Fl. 12, the same amount that a skilled printer involved in setting the *Theatrum* could expect to earn in a whole month.[24] Rather than deterring buyers, the price of this new geographical text and the time and skill invested in its printing clearly became one of its most attractive features. One satisfied customer wrote to Ortelius informing him that 'the maps are very distinctly printed, and the paper is beautiful and white'.[25] Yet it was Mercator's praise for the atlas, already briefly discussed in Chapter One, which was indicative of the extent to which Ortelius had managed to unite intellectual innovation and comprehensiveness with an astute eye to the commodification of the geographical image. In his letter to Ortelius written upon receipt of a copy of the *Theatrum* in 1570, Mercator praised him for having created a geographical artefact which was not only comprehensive but could be 'kept in a small place, and even carried about wherever we wish'. Mercator was convinced that 'this work of yours will always remain saleable, whatever maps may in the course of time be reprinted by others'.[26]

However, Mercator's praise was based on a mistaken assumption that, like his own world map, the *Theatrum* would stand as a static, historically monolithic representation of geographical knowledge. In fact, Ortelius' plans for the *Theatrum* were even more ambitious than Mercator had envisaged, and represented an entirely new approach to the production and distribution of such knowledge. Ortelius had created a printed format whereby the skeleton structure of the *Theatrum* could remain in place, but increasingly incorporate a potentially endless variety of new maps and information. By the end of 1570 another four editions had been printed, with minor changes. By 1571 a Dutch edition had appeared, and by 1573 a new Latin edition was printed, this time with new maps. By 1579 the *Theatrum* had expanded to include ninety-seven map sheets, incorporating maps of the classical world which firmly placed the classical geography of Ptolemy as supplemental to the contemporary geographical images produced by Ortelius and his network of geographers. Year by year the *Theatrum* grew in size and diversity until by 1612 there were no fewer than forty-two editions in circulation, continually updated and expanded in Dutch, German, French, Spanish, English and Italian translations, as well as the original Latin. The brilliance of Ortelius' *Theatrum*, as Mercator partly realized, lay in its ability to condense the diversity of both classical and contemporary geographical knowledge into a single volume, from which its owners could extract the information they required in an instant, rather than consulting cumbersome wall maps which then had to be cross-referenced to a mass of travel reports, geographical descriptions and diplomatic documents. Another admiring owner of a copy of the *Theatrum* wrote to Ortelius in 1579 exclaiming:

You compress the immense structure of land and sea into a narrow space, and have made the earth portable, which a great many people assert to be immoveable.[27]

Not only had Ortelius provided the consummate geographical artefact which insulated its readers from the deracinating effects purportedly suffered by the early Portuguese voyagers. He had also created the ideal medium through which to endlessly mediate the countless voyages, travels, encounters, exchanges and discoveries which the field of sixteenth-century geography attempted to record comprehensively. In his 'Address to the Reader' he defined geography, and implicitly the function of his *Theatrum*, as 'The eye of History'.[28] Yet in continuing this visual analogy, Ortelius conceded the extent to which such a vision was an endlessly mediated process:

This so necessary a knowledge of Geography, as many worthy and learned men have testified, may very easily be learn'd out of Geographicall Chartes or Mappes. And when we have acquainted our selves somewhat with the use of these Tables or Mappes, or have attained thereby to some reasonable knowledge of Geography, whatsoever we shall reade, these Chartes being placed, *as it were certaine glasses before our eyes*, will the longer be kept in the memory, and make the deeper impression in us: by which meanes it commeth to passe, that now we do seeme to perceive some fruit of that which we have read.[29]

Ortelius' rhetoric aimed to ensure that the 'glass' placed before the majority of the literate map-buying public of the late sixteenth century was the *Theatrum*, and that the endless representational mediations which made up this eye on the world were only ultimately ratified by the authority of the maker of the glass: the geographer. Encompassing contemporary topography, discovery and political administration as well as classical geography, Ortelius' *Theatrum* not only shaped the future direction of geographical representation, but it also undertook to define the historical shape of geography as an intellectual discipline in its own right, thus finally freeing the field from the over-determined influence of Ptolemy and the presuppositions of classical geography.

The result was a further intensification of the professionalization of the geographer and the commodification of his products. Ortelius emphasized the value of his *Theatrum* in the closing words of his 'Address to the Reader', where he announced that:

Wherefore the students of Geography shall have here, in the Authors thus named in order, and in the Catalogue of the Authors of the Geographicall Tables or Mappes, which we have set before this our worke, and lastly in these Tables themselves, *a certain shoppe*, as it were, furnished with all kinde of instruments necessarily required in such like businesse: out of which, if peradventure there may seeme anything wanting, in his judgement, either to the finishing of any Booke of that argument, or more fuller descriptions of any Countreys what-soever, very easily, or in deed without any labour at all he may see, from whence it may by and by be fetched.[30]

In comparing the *Theatrum* to a shop, rather than stressing its value as a precious intellectual commodity, Ortelius naturalized its presence as an indispensable part of the social and political life of the élites of late sixteenth-century Europe. In marketing it in this way, he ensured that his atlas would not be perceived as a politically partial geographical object, in the way that Ribeiro's world map of 1529 had come to be asso-ciated with Charles V's claim to the Moluccas. Like Mercator, Ortelius maximized not only the political neutrality of his work, but also its com-mercial success by stressing its utility to diplomats, merchants and scholars of whatever political persuasion. In his 'Address' he therefore stated his hope that 'every student would affoord [the *Theatrum*] a place in his Library, amongst the rest of his bookes'.[31] The *Theatrum* was to take its place in a textual universe of learning, expunged of the deleterious effects of travel. Diplomats and statesmen could henceforth shape the territorial and commercial contours of the world, regardless of the effects which such decisions would have beyond the confines of their libraries. Whilst intel-lectually the *Theatrum* became a 'library without walls'[32] as it relentlessly expanded the parameters of geographical knowledge edition after edition, politically it allowed politicians to circumscribe the limits of the expanding world and imaginatively carve it up in line with their own particular poli-tical preoccupations.

The very fact that the *Theatrum* so loudly proclaimed its neutrality only enhanced its status as a politically significant object. Like Mercator, Ortelius attracted imperial patronage which enhanced the social esteem in which his work was held, and this led to attempts by its politically powerful patrons to put their own mark, very literally, on such an appar-ently objective geographical product. Yet even this overt political intervention into Ortelius' mapmaking was turned to the geographer's commercial and financial benefit. In 1571 he forwarded a copy of the *Thea-trum* for presentation to Charles V's son, the Spanish king Philip II. The copy was initially inspected by Cardinal Espinosa, chief minister to Philip, who searched in vain for his birthplace of Martimuñoz on the map of Spain printed in the *Theatrum*. Espinosa's aide, Hieronymus de Rhoda, swiftly wrote back to Ortelius informing him that:

Cardinal [Espinosa] writes me from Spain that he regrets to find his native town Martimuñoz, omitted in the map of Spain in your 'Theatrum' and asks me to send him a coloured copy of it with the said name inserted. Therefore remove, if possible, the name of Palacuelos and insert Martimuñoz in its place. When this is done let some copies of the map of Spain be printed to satisfy the

Cardinal. And as a fleet will sail to Spain, with the first favourable wind, send by it two coloured copies of the 'Theatrum' with the altered name; bound in leather and gilt. Aries Montanus will pay the necessary expense.[33]

Subsequent editions of the *Theatrum* show the erasure of Palacuelos and the insertion of Martimuñoz (illus. 50). Whilst there is no record of Ortelius' financial renumeration in undertaking this cartographical cooking of the books, it did have one particularly striking effect. Later on in the same year Montanus was ordered to confer upon Ortelius the title of Geographer Royal of His Majesty Philip II.[34] This prestigious award (to some extent a historical extension of Mercator's own geographical services to the House of Habsburg) was clearly a reward for Ortelius' swift reprinting of his atlas to suit the tastes of Cardinal Espinosa, one of Philip's closest political confidants.

This remarkable incident exemplifies the extent to which Ortelius' *Theatrum* had become established as the most prized geographical possession of the sixteenth century. But perhaps even more significantly, it stresses the important shift in the power relations which increasingly characterized the production of later sixteenth-century geography. In the 1520s the Habsburg authorities had been able to dictate the terms of geographical representation, which had effectively portrayed a world picture that reflected a range of commercial and territorial objectives in their own image. By the 1570s the same authorities were to be found petitioning increasingly powerful and authoritative geographers like Mercator and Ortelius for political concessions in the way they depicted the world. The commercial considerations which had created the conditions for the cartography of the likes of Ribeiro and the Reinel brothers in the 1530s had ultimately given the geographer an unprecedented degree of intellectual and political autonomy, which he proceeded to exercise in the social and commercial dissemination of his printed maps and globes. Espinosa's intervention in the production of the *Theatrum* stressed that political partiality in the production of socially influential maps and globes could be purchased: but at a price which was no longer indexed to political loyalty but instead based its value on those of the market-place. By 1588 the Plantin press responsible for printing Ortelius' *Theatrum* took full commercial advantage of the imperial patronage the work had attracted by issuing a luxurious copy of the first Spanish edition and dedicating it to Prince Asturias, the heir to the Spanish throne. This opulently illuminated and elaborately bound copy cost a staggering Fl. 90.[35] In taking care to accede to the demands of figures such as Espinosa, Plantin and Ortelius ensured that, like Mercator, they reaped the maximum intellectual and financial rewards available from tacitly acknowledging the patronage of their Habsburg paymasters.

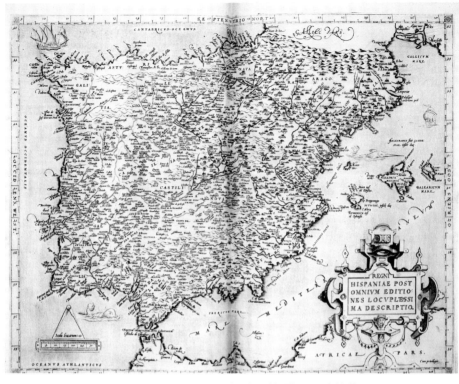

50 Abraham Ortelius, map of Spain, copperplate engraving from his *Theatrum Orbis Terrarum*, Antwerp, 1573. British Library, London.

The End of the Line?

The one thing that the dispute over the Moluccas had established once and for all was the social and political value of geography, as both a logistical tool which facilitated the activities of merchants and diplomats and a politically valuable resource in effectively laying claim to commercially prized territories overseas. However, in allowing geographers a degree of intellectual freedom and creative autonomy to create a more comprehensive and standardized representation of the world, political patrons of the discipline saw their authority to dictate the political contours of geographical representation increasingly eroded. The commercial and intellectual possibilities which the printing press created allowed geographers like Mercator and Ortelius to appreciate that their output would be more commercially successful and intellectually respected if it distanced itself from the politically partial geography which had been produced around the dispute over the Moluccas. By strategically drawing on their imperial patronage, both geographers inexorably redefined the social status of early modern geography as a discipline defined predominantly by its claims to intellectual objectivity and the demands of the market, rather than the demands of an imperial crown or a politically influential patron. It was this shift which finally saw the emergence of a more recognizably 'modern' discipline, which eschewed the principles of political speculation and the mysterious aura of the geographer which had defined late fifteenth- and sixteenth-century geography. Instead, the defining principles of the discipline in the seventeenth century became objectivity and accuracy, as geography gradually transformed itself from a mysterious art into a scientifically exact field of political and commercial administration. This in no way saw the development of a 'true' science of geographical representation; if anything, the representational procedures which cartography developed produced even more sophisticated techniques with which to offer highly partial and ideological representations of the world.[36] However, this shift towards claims to transparency and objectivity did signal the death of early modern geography and the birth of a definably 'modern' discipline, and it is with this shift that this particular study concludes.

6 Conclusion

Throughout the course of this book I have argued for the importance of what could be called the presence of the East in defining early modern European understanding of travel, commerce and geography. It was from this 'eastern horizon' that sixteenth-century Europe increasingly came to define its own social and geographical identity, as it gradually redefined itself in opposition to the old classical world depicted in Ptolemy's *Geographia*, and ultimately turned its back on this Old World to embrace an increasingly predominant Atlantic World sketched out in Mercator's 1569 map. It therefore seems appropriate to conclude this study by once more invoking a map which signals the end of early modern geography, but at the same time retains traces of the continued presence of the East within the shaping of early European identity.

Petrus Plancius' map of the Moluccas Islands, dated 1594 and entitled *Insulae Moluccae* (illus. 51), was symptomatic of the new social forces which increasingly came to define the field of seventeenth-century geography, whilst also implicitly looking backwards to the disputes of the first half of the sixteenth century which ultimately created this shift in the field of geographical representation. Plancius' clinical depiction of the spice-producing islands which had caused so much controversy between the crowns of Castile and Portugal is a vivid example of the changing political and commercial landscape of the early modern world. Unlike Ribeiro's world map of 1529, or even Mercator's map of 1569, Plancius' map had not been produced under the auspices of imperial patronage. Instead its careful topographical and hydrographical information, and exact reproduction of commercially valuable commodities like nutmeg, cloves and sandalwood, were designed to aid the commercial initiatives of the new *Compagnie van Verre*, a Dutch company formed in Amsterdam in 1594. It was this company which was to form the basis of the seventeenth-century joint-stock company the *Vereenigde Oost-Indische Compagnie*, otherwise known as the Dutch East India Company. The map therefore reflected one of the most significant political and commercial shifts to affect early modern Europe at the turn of the sixteenth

51 Petrus Plancius, *Insulae Moluccae*, engraving, Antwerp, 1594. British Library, London.

century. This was the gradual erosion of the imperial authority of the Habsburg and Portuguese crowns in Southeast Asia, and their replacement by the aggressively mercantile power and authority of the Dutch and English East India companies. Privately funded and organized by financiers and merchants, these companies represented a distinct development in the commercial ethos which had characterized the activities of the empires of the Portuguese and the Habsburgs in Southeast Asia throughout the sixteenth century. Rather than speculating on commercial ventures to far-flung places, these new companies sought to incorporate the entire trading network of the Indian and Pacific Oceans into one enormous monopoly. A string of commercial offices was established between northern Europe and Southeast Asia to co-ordinate the flow of merchandise along the shipping lanes which connected these two points on the map of the world.

The consequences of this shift in political and commercial organization was substantial. Not only did it lead to an intensification and regulation of the range and size of commodities which moved back and forth between Europe and Asia, but it also led to the demand for ever more precise and objective geographical products. Increasingly, the most up-to-date maps, charts and globes were produced not under the auspices of imperial courts, but to the requirements of commercial companies such as the Dutch East India Company and its English rival, the East India Company. The most respected geographers working in the late sixteenth century owed their allegiance not to their crowns, but to the joint-stock companies which based themselves in the metropolitan centres of early modern Europe. Upon the formation of the English East India Company in 1601, the English geographer and travel writer Richard Hakluyt was employed to provide maps of the East Indies to facilitate its commercial activities. The minutes of the company's meeting of 29 January 1601 recorded that:

Mr. Hacklett, the historiographer of the voyages of the East Indies, being here before the Committees and having read unto them out of his notes and books divers instructions for provisions of Jewels, was required to set down in writing a note of the principal places in the East Indies where Trade is to be had, to the end the same may be used for the better instruction of our factors in the said voyage.[1]

Less than a month later the records of the company noted that:

There is given to Mr. Hackett [*sic*] by the assent of this assembly for his travails taken in instructions and advices touching the preparing of the voyage and for his former advices in setting the voyages in hand the last year the sum of ten pounds, and thirty shillings for three maps by him provided and debited to the company.[2]

Like Plancius, Hakluyt became an informal employee of a joint-stock company that was concerned not with partial geography which inflated its territorial claims to far-flung territories, but with accurate geographical descriptions of the sea routes and topography of the areas it specified as crucial in expanding their commercial presence. By the 1620s the Dutch East India Company had created a hydrographic office in Amsterdam which sought to co-ordinate the systematic mapping of the territories across which the company established its theatre of commercial operations.[3] So sophisticated had this mapping process become that the offices in Amsterdam quickly established one in Batavia, to create new maps and charts of the Indonesian region which were forwarded to Amsterdam for incorporation into the charts and globes produced for the exclusive use of the pilots and administrators of the company.

The effect which this new process of mapping had upon the field of geography was gradual but profound. In attempting to systematically map every inch of distant territory through local surveys, the geographical and hydrographical practices of the new commercial companies spelt the end of the speculative geography of the early sixteenth century whose principles have formed the basis of this particular study. The maps and globes of earlier geographers like Behaim and Ribeiro had deliberately exploited the partial and often conflictual accounts of distant territories to create politically and commercially compelling 'imaginative geographies' which convinced backers with a mixture of aesthetic beauty and loudly proclaimed scholarly wisdom. The practical difficulties of physically reaching remote places allowed the geographer to utilize his conjectural knowledge of geographically distant phenomena to invest both his products and his social status with an aura of prestige and awe at his ability to represent, almost magically, such distant and inaccessible places.

The processes of mapping that developed throughout the early seventeenth century gradually eroded this perception of the geographer and his products. The joint-stock companies were not interested in speculating on commercially valuable and geographically distant territories, or in commissioning studies which inflated their own commercial and political importance. What mattered to them was the accurate, objective and systematic mapping of strategic territory. As a result, geographical specialists like Plancius and Hakluyt, initially self-styled savants on the lines of Behaim and Ribeiro, were absorbed into the institutional structures of organizations like the East India companies, which provided ample opportunities for them to exercise their wisdom in exchange for a specific position and salary. The new model of geographical production was exemplified in the monumental globes and atlases produced in Holland throughout the seventeenth century, which have been most famously depicted in the paintings of Jan Vermeer. The *Atlas Maior* of the great Dutch geographer Joan Blaeu, published in 1662, was symptomatic of this changing perception of geography (illus. 52). Blaeu's was the first geographical atlas to finally surpass Ortelius' *Theatrum* in its scale and comprehensiveness. Comprising twelve volumes with 3,000 pages of text and approximately 600 maps, the *Atlas* was a massive intellectual and commercial undertaking. Indirectly its production had been financed by the Dutch East India Company, who had appointed Blaeu as its official mapmaker in 1638.[4] Blaeu therefore exemplified the new model of the geographical specialist, which had been anticipated by Ortelius in his own skilful marketing of his *Theatrum*. Employed by the company and utilizing its growing body of carefully collated charts and geographical information, Blaeu represented a new type of geographer who eschewed speculation in the pursuit of the construction of a globally comprehensive geography.

Blaeu's career in relation to the Dutch East India Company was in stark contrast to the careers of the earlier geographers who have been examined throughout the course of this book. Organizations like the Dutch and English East India companies were too large and impersonal to accommodate the entrepreneurial activities of a geographical figure like Behaim who, in the absence of an organized commercial company at the end of the fifteenth century, acted as not only a geographer but also a traveller, merchant, mathematician and globemaker. The companies gradually separated out all these jobs into specific offices and administrative duties, thus denying the geographer the opportunity to move across a range of disciplines which had previously made maps and globes so compelling in their ability to permeate so many corners of early modern life. The incorporation within the companies of the most up-to-date geographical material marked the decline of the map as a skilfully marketed and coveted material object, capable of crossing political and cultural boundaries. The

NOVA ET ACCVRATISSIMA TOTIVS

52 Joan Blaeu, world map, engraving from his *Atlas Maior*, Amsterdam, 1662.
British Library, London.

RVM ORBIS TABVLA. *Auctore* IOANNE BLAEV.

irony was that, in its emergence along the trade routes via the Cape of Good Hope and deep into Southeast Asia throughout the late fifteenth and sixteenth centuries, the discipline of geography had assisted in the successful establishment of a commercial network which, by the beginning of the seventeenth century, saw the geographer reduced to the role of skilled administrator, whilst his maps increasingly functioned as hydrographical and topographical tools, useful commodities with which to approach the business of trade and ultimately, as the seventeenth century progressed, colonization.

Late fifteenth- and sixteenth-century geography had only ever really been based on the ability to speculate and conjecture on the imaginative possession of distant territories. The insurmountable difficulties of actually implementing the claims to global authority and possession made by the maps and globes of geographers like Ribeiro and Frisius did not, however, make them any less effective and valuable in the eyes of the merchants and diplomats who used them. Precisely because of the impossibility of seriously laying claim to global possession, the map and the globe became the most powerful instruments through which to construct a rhetoric of claims to commercial and political power and authority, as Charles V showed in the dispute over the Moluccas. They were intensely valuable mediators in the construction of the social life of the early modern period. Their commercial and political value only changed at the point at which the field of geography assimilated itself within the administrative and scientific structures of seventeenth-century European society. This assimilation increasingly divested the map of its power to mediate the distant and mysterious limits of the early modern world. Rather than being a source of wonder and mystery, the geographical object, be it map or globe, increasingly laid claim to its presence as a studiedly transparent image of an increasingly known world. At this moment early modern geography lost its fascinating ability to entwine commerce, finance, travel and diplomacy in its dense and often fragmented images of the world. It was at this point that a certain ethos which defined the mapping of the early modern world came to an end.

References

1 *Introduction*

1 B. Diffie and G. Winius, eds, *Foundations of the Portuguese Empire* (Minneapolis, 1977), p. 185.

2 Annemarie Jordan Gschwend, 'In the tradition of princely collections: curiosities and exotica in the *Kunstkammer* of Catherine of Austria', *Bulletin of the Society for Renaissance Studies*, XIII, 1 (1995), pp. 1–9.

3 See for example D. W. Waters, 'Science and the techniques of navigation in the Renaissance' in *Art, Science and History in the Renaissance*, ed. C. Singleton (Baltimore, 1967), pp. 189–237.

4 Cited in Peter van der Krogt, *Globi Neerlandici. The Production of Globes in the Low Countries* (Utrecht, 1993), p. 76.

5 John Dee, *Preface to Euclid's 'Elements of Geometrie'* (London, 1570), sig.a.iiij.

6 On the concept of 'Renaissance Man', see Jacob Burckhardt's highly influential study, *The Civilisation of the Renaissance in Italy* (London, 1892). Whilst Burckhardt's thesis has been comprehensively interrogated more recently, many critics still remain ambivalent concerning Burckhardt's legacy; see for instance Stephen Greenblatt, *Renaissance Self-Fashioning: From More to Shakespeare* (Chicago, 1980). More recently see Margreta de Grazia 'The ideology of superfluous things: *King Lear* as period piece', in *Subject and Object in Renaissance Culture*, ed. Margreta de Grazia *et al.* (Cambridge, 1996), pp. 17–42.

7 Donald Lach, *Asia in the Making of Europe: A Century of Wonder*, I, Book 2 (Chicago, 1970), pp. 95–98.

8 *Ibid.*, p. 55. On dyes and dyeing see Grete de Francesco, 'The Venetian silk industry', *Ciba Review* 29 (1940), pp. 1027–35.

9 J. H. Parry, *The Age of Reconnaissance: Discovery, Exploration and Settlement 1450–1650* (London, 1963), pp. 132–46. See also the opening sections of Clarence Henry Haring, *Trade and Navigation Between Spain and the Indies* (Massachusetts, 1918) on the *Casa* and the ways in which the Spanish developed a similar model in Seville.

10 I am grateful to Professor Denis Cosgrove for providing me with this translation from the Italian.

11 Cited in Armando Cortesão and Avelino Teixeira da Mota, *Portugaliae Monumenta Cartographica* (Coimbra, 1958–63), I, p. 12.

12 *Ibid.*, pp. 12–13.

13 Tony Campbell, 'Portolan charts from the late thirteenth century to 1500', in *The History of Cartography: Cartography in Prehistoric, Ancient, and Mediaeval Europe and the Mediterranean*, I, eds J. B. Harley and D. Woodward (Chicago, 1987), pp. 371–463: p. 437.

14 Cited in D. B. Quinn, ed., *The Hakluyt Handbook* (London, 1974), I, p. 305.

15 See Burckhardt, *The Civilisation of the Renaissance*; more recently John Hale, *The Civilisation of Europe in the Renaissance* (London, 1994), has explicitly reworked Burckhardt's model of the Renaissance. Like Burckhardt, Hale's text almost

completely elides Arab and Ottoman influences in Renaissance culture. For a critique of this 'Orientalist' tendency (although concerned more generally with late English Renaissance culture), see Kim Hall, *Things of Darkness: Economies of Race and Gender in Early Modern England* (Cornell, 1995).

16 E. G. Ravenstein, ed., *A Journal of the First Voyage of Vasco da Gama* (London, 1898), p. 60.

17 Helder Macedo, 'Recognizing the unknown: perceptions in the age of European expansion', *Portuguese Studies*, 8 (1992), pp. 130–36: p. 135.

18 Jonathan Lanman, 'The religious symbolism of the T in T-O maps', *Cartographica*, 18, 4 (1991), pp. 18–22.

19 Mary Helms, *Ulysses' Sail: An Ethnographic Odyssey of Power, Knowledge, and Geographical Distance* (Princeton, 1988), pp. 212–16.

20 See J. B. Friedman, 'Cultural conflicts in mediaeval world maps', in *Implicit Understandings. Observing, Reporting, and Reflecting on the Encounters Between Europeans and Other Peoples in the Early Modern Era*, ed. Stuart B. Schwartz (Cambridge, 1994), pp. 64–95. For a fascinating account of the ways in which such travels began to dissolve the coordinates of the T-O symbology, see Stephen Greenblatt, *Marvelous Possessions: The Wonder of the New World* (Oxford, 1991), pp. 26–51.

21 On Fra Mauro see R. Almagia and T. Gasparrini, *Il Mappamondo di Fra Mauro* (Florence, 1956).

22 Cited in G. R. Crone, *Maps and their Makers* (London, 1953), p. 58.

23 Peter Barber, 'Visual encyclopaedias: the Hereford and other Mappae Mundi', *The Map Collector*, 48 (1989), pp. 2–8.

24 R. A. Skelton, 'A contract for world maps at Barcelona, 1399–1400', *Imago Mundi*, 22 (1968), pp. 107–13.

25 On the transmission of Ptolemy's text, see O. A. W. Dilke's studies, *Greek and Roman Maps* (London, 1985), and 'Cartography in the Byzantine empire', in *The History of Cartography: Cartography in Prehistoric, Ancient and Mediaeval Europe and the Mediterranean*, I, eds J. B. Harley and D. Woodward (Chicago, 1987), pp. 258–75. See also Noel Swerdlow, 'The recovery of the exact sciences of antiquity: mathematics, astronomy, geography', in *Rome Reborn: The Vatican Library and Renaissance Culture*, ed. Anthony Grafton (Washington, 1993), pp. 125–67.

26 On this shift see David Woodward, 'Maps and the rationalization of geographic space', in *Circa 1492: Art in the Age of Exploration*, ed. Jay Levenson (New Haven, 1991), pp. 83–88.

27 *Ibid.*, pp. 218–224.

28 Edward Said, *Orientalism* (London, 1978).

29 For an excellent account of the early print history of Ptolemy's text, see Tony Campbell, *The Earliest Printed Maps* (London, 1987), pp. 122–38.

30 *Ibid.*, p. 130.

31 On the impact of print on early modern culture, see Elizabeth Eisenstein, *The Printing Press as an Agent of Change: Communications and Cultural Transformations in Early Modern Europe*, 2 vols (Cambridge, 1979); Lucien Febvre and and Henri-Jean Martin, *The Coming of the Book: The Impact of Printing 1450–1800* (London, 1976); on the specifications of map production in relation to the new technology of print, see R. A. Skelton, 'The early map printer and his problems', *Penrose Annual*, 57 (1964), pp. 171–87.

32 William Ivins, *Prints and Visual Communications* (New York, 1969), pp. 1–50.

33 Campbell, *Earliest Printed Maps*, p. 1.

34 Abraham Ortelius, 'Address to the Reader', *Theatrum Orbis Terrarum* (London, 1606), unpaginated.

35 On the enormous range of scholars and geographers Ortelius cited in his *Theatrum*, see Robert Karrow's excellent *Mapmakers of the Sixteenth Century and their Maps: Bio-Bibliographies of the Cartographers of Abraham Ortelius, 1570* (Chicago, 1993).

36 Cited in J. H. Hessels, ed., *Ortelii Epistulae* (Oxford, 1887), no. 32.

37 On the rapid changes to later world maps based on Ptolemy, see Miriam Usher Chrisman, *Lay Culture, Learned Culture: Books and Social Change in Strasbourg, 1480–1599* (New Haven, 1982), pp. 136–38.

38 On the historical development of the portolan chart, see Campbell, 'Portolan charts', pp. 371–463.

39 See for instance Lloyd Brown, *The Story of Maps* (New York, 1949), Ronald Tooley, *Maps and Map-Makers* (London, 1949), and Norman Thrower, *Maps and Civilization: Cartography in Culture and Society* (Chicago, 1996). Probably the best recent survey dealing with printed maps of the period is Rodney Shirley, *The Mapping of the World: Early Printed Maps, 1472–1700* (London, 1983).

40 Plancius was specifically employed to 'make a special register of all the places in the East India, where trade may be looked for'. Cited in F. C. Wieder, *Monumenta Cartographica*, I (The Hague, 1925–33), p. 44. On Plancius, see F. C. Wieder, *De Oude Weg Naar Indië Ons De Kaap* (Amsterdam, 1915). On the early development of the specific nature of Dutch commercial activity, see Jonathan Israel, *Dutch Primacy in World Trade, 1585–1740* (Oxford, 1989).

2 *An Empire Built on Water: The Cartography of the Early Portuguese Discoveries*

1 John Hale allocates a mere two pages to the impact of the Portuguese voyages upon the culture of early modern Europe. Hale, *Civilization of Europe*, pp. 315–16.

2 J. H. Parry, *Europe and a Wider World, 1415–1715* (London, 1949), and *The Age of Reconnaissance: Discovery, Exploration and Settlement* (London, 1973); Boies Penrose, *Travel and Discovery in the Renaissance* (Massachusetts, 1952).

3 Cited in C. R Boxer, *The Portuguese Seaborne Empire, 1415–1825* (New York, 1969), p. 87.

4 The standard study of Portuguese cartography remains Armando Cortesão and Avelino Teixeira da Mota, *Portugaliae monumenta Cartographica*, 6 vols (Coimbra, 1958–63) but see also Avelino Teixeira da Mota, 'Evolução dos roteiros portugueses durante o século XVI', *Revista da Universidade de Coimbra*, 24 (1969), pp. 1–32, and Alfredo Pinheiro Marques, *Origem e desenvolvimento da cartografia portuguesa na época dos descobrimentos* (Lisbon, 1987).

5 On the changing perceptions of the status of geography and cosmography in later fifteenth-century Portugal, see Patricia Seed, *Ceremonies of Possession in Europe's Conquest of the New World 1492–1640* (Cambridge, 1995), pp. 116–20, and David Turnbull, 'Cartography and science in early modern Europe: mapping the construction of knowledge spaces', *Imago Mundi*, 48 (1996), pp. 5–24, for an account of the ways in which the Portuguese crown attempted to create what Turnbull calls a centralized 'knowledge space' within which to assemble and standardize cartographic information.

6 Cited in Cortesão and da Mota, *Portugaliae monumenta Cartographica*, I, p. 12.

7 Cited in Francis M. Rogers, ed., *The Travels of the Infante Dom Pedro of Portugal* (Massachusetts, 1961), p. 48.

8 See E. G. R. Taylor, *The Haven-finding Art: A History of Navigation from Odysseus to Captain Cook* (London, 1956), and for a specific account of the Portuguese responses to these problems see Mota, 'Evolução'.

9 On the history and development of Portuguese navigation, see Joaquim Bensaúde, *Histoire de la science nautique portugaise a l'époque des grandes découvertes*, 7 vols (Lisbon, 1914–24), and the work of Luís de Albuquerque, *Curso de História da Nautica* (Coimbra, 1972), and *Historia de la navegación portuguesa* (Madrid, 1991).

10 Cited in Diffie and Winius, *Foundations of the Portuguese Empire*, p. 138.

11 Armando Cortesão, *History of Portuguese Cartography*, II (Coimbra, 1969), p. 216.

12 On this perception of the Portuguese 'discoveries', see Seed, *Ceremonies of Possession*, pp. 100–48.

13 Gomes Eannes de Azurara, *The Chronicle of the Discovery and Conquest of Guinea* (London, 1896), p. 29. For the Portuguese text, see José de Braganza, ed., *Crónica do descobrimento e conquista da Guiné*, 2 vols (Oporto, 1937).

14 Cited in John Blake, ed., *European Beginnings in West Africa, 1457–1578* (London, 1937), pp. 67–8.

15 Cited in G. R. Crone, ed., *The Voyages of Cadamosto and Other Documents on Western Africa in the Second Half of the Fifteenth Century* (London, 1937), p. 51. Emphasis added. For the Portuguese text, see João Franco Machado and Damião Peres, eds, *Viagens de Lúis de Cadamosto e Pedro Sintra* (Lisbon, 1948).

16 On the development of the myth of Henry as symptomatic of a peculiarly late eighteenth- and nineteenth-century fusion of national and imperial self-definition in Portuguese historiography, see Cândido Lusitano, *Vida do Infante D. Henrique* (Lisbon, 1758), Fortunato de Almeida, *O Infante de Sagres* (Oporto, 1894). For English reworkings of the myth, see Charles Raymond Beazley, *Prince Henry the Navigator* (New York, 1923), and Elaine Sanceau, *Henry the Navigator* (New York, 1947). For a critique of the contemporary colonial politics which underpinned such accounts, see A. Guimaràes, *Uma corrente do colonialismo Português: a Sociedade de Geografia de Lisboa, 1875–1895* (Lisbon, 1984). For critiques of Henry and his mythic learning, see Duarte Leite, *Descobrimentos portuguêses*, 2 vols (Lisbon, 1958–60), I, pp. 163–70, and more recently W. G. L. Randles, 'The alleged nautical school founded in the fifteenth century at Sagres by Prince Henry of Portugal, called "The Navigator" ', *Imago Mundi*, 45 (1993), pp. 20–28.

17 Azurara, *Chronicle of Guinea*, p. 248.

18 Cited in Diffie and Winius, *Foundations of the Portuguese Empire*, pp. 303–4.

19 See Vitorino Magalhães Godinho's studies, *A economia dos descobrimentos henriquinos* (Lisbon, 1962), and *Os descobrimentos e a economia mundial*, 2 vols (Lisbon, 1965–71).

20 On these initiatives see Crone, ed., *The Voyages of Cadamosto*.

21 Diffie and Winius, *Foundations of the Portuguese Empire*, pp. 151–59.

22 Felipe Fernández-Armesto, *Before Columbus: Exploration and Colonisation from the Mediterranean to the Atlantic, 1229–1492* (London, 1987), p. 200.

23 Cited in Francis M. Rogers, ed., *The Obedience of a King of Portugal* (Minneapolis, 1958), p. 48.

24 For what remains one of the best accounts of Behaim and the production of his globe, see E. G. Ravenstein, *Martin Behaim: His Life and His Globe* (London, 1908), but also see Damião Peres, *História dos descobrimentos portuguêses* (Oporto, 1960), pp. 277–86, and more recently the essays in *Focus Behaim-Globus* (Nuremberg, 1991).

25 Cited in Ravenstein, *Martin Behaim*, pp. 71–72.

26 *Ibid.*, p. 87.

27 Cited in Diffie and Winius, *Foundations of the Portuguese Empire*, p. 171.

28 Frances Gardiner Davenport, ed., *European Treatises on the History of the United States and its Dependencies*, I (Washington, 1967), p. 95.

29 Cited in Ravenstein, ed., *A Journal of the First Voyage of Vasco da Gama*, p. 60. For the Portuguese text, see António de Magalhães Basto, ed., *Diário da viagem de Vasco da Gama* (Oporto, 1945).

30 Ravenstein, *Journal*, p. 60.

31 *Ibid.*, p. 114.

32 Cited in Donald Weinstein, ed., *Ambassador From Venice: Pietro Pasqualigo in Lisbon, 1501* (Minneapolis, 1960), p. 46.

33 *Ibid.*, p. 46.

34 *Ibid.*, pp. 31–32.

35 *Ibid.*, pp. 29–30. For a discussion of the commercial contexts of changes to the spice

trade, see Wilhelm von Heyd, *Histoire de commerce du Levant au Moyen Age* (Leipzig, 1885–86), II.

36 Cited in Weinstein, *Ambassador From Venice*, p. 30.

37 Palmira Brummett, *Ottoman Seapower and Levantine Diplomacy in the Age of Discovery* (New York, 1994), p. 45. For primary documentation on Portuguese involvement in the Indian Ocean, see Gaspar de Corrêa, ed., *Lendas da Índia*, 8 vols (Lisbon, 1856–66).

38 Cited in W. B. Greenlee, ed., *The Voyage of Pedro Alvares Cabral to Brazil and India* (London, 1937), p. 123.

39 *Ibid.*, p. 124.

40 Avelino Teixeira da Mota, 'Some notes on the organization of hydrographical services in Portugal before the nineteenth century', *Imago Mundi*, 28 (1976), pp. 51–60: p. 53.

41 Giovanni Matteo Contarini, *A Map of the World, Designed by Giovanni Matteo Contarini* (London, 1926), p. 8.

42 William Cunningham, *The Cosmographicall Glasse* (London, 1555), p. 120.

43 Cited in Francis Maddison, 'A consequence of discovery: astronomical navigation in fifteenth-century Portugal', in *Studies in the Portuguese Discoveries I*, eds T. F. Earle and S. Parkinson (Warminster, 1992), pp. 71–110: pp. 71–72. On da Gama's utilization of local knowledge, see José Brocahdo, *O piloto Árabe de Vasco da Gama* (Lisbon, 1959).

44 Cited in Sergio Pacifici, ed., *Copy of a Letter of the King of Portugal Sent to the King of Castile Concerning the Voyage and Success of India* (Minneapolis, 1955), p. 13.

45 Cited in Crone, *Maps and their Makers*, p. 92.

46 *Ibid.*, p. 87.

47 For just one example of the ways in which Portuguese historiography has started to redefine its perceptions of Portugal's early modern maritime activities, see L. Maduerira, 'The discreet seductiveness of the crumbling empire', *Luso-Brazilian Review*, 32, 1 (1995), pp. 17–29. I am grateful to James Siddaway for bringing this article to my attention.

3 *Disorienting the East: The Geography of the Ottoman Empire*

1 On Berlinghieri, see R. A. Skelton, ed., *Francesco Berlinghieri: Geographia* (Amsterdam, 1966).

2 Campbell, *Earliest Printed Maps*, pp. 133–35. For other contemporary examples of tentative revisions of Ptolemy's maps, see David Woodward, 'Medieval *Mappaemundi*', in *The History of Cartography: Cartography in Prehistoric, Ancient and Mediaeval Europe and the Mediterranean*, I, eds J. B. Harley and D. Woodward (Chicago, 1987), pp. 286–370: p. 316.

3 Cited in Kemal Özdemir, *Ottoman Nautical Charts and the Atlas of 'Ali Mācār Reis* (Istanbul, 1992), p. 52.

4 See Rifaat A. Abou-el-Haj, 'The formal closure of the Ottoman frontier in Europe, 1699–1703', *Journal of the American Oriental Society*, 89 (1969), pp. 467–75.

5 See C. F. Black, ed., *et al.*, *Atlas of the Renaissance* (Amsterdam, 1993), p. 134.

6 For a classic account of the 'horror' of the fall of Constantinople and its immediate aftermath, see Robert Schwoebel, *The Shadow of the Crescent: The Renaissance Image of the Turk, 1453–1517* (Nieuwkoop, 1957).

7 Cited in Gülru Necipoğlu, *Architecture, Ceremonial and Power: The Topkapi Palace in the Fifteenth and Sixteenth Centuries* (Massachusetts, 1991), p. 12. On the ongoing comparisons between the Ottoman sultans and Alexander, which extended right through to the reign of Süleyman the Magnificent, see Necipoğlu, 'Süleyman the Magnificent and the representation of power in the context of Ottoman-Habsburg-Papal rivalry', *Art Bulletin*, 71 (1989), pp. 401–27.

8 Cited in Necipoğlu, *Architecture, Ceremonial and Power*, p. 12.

9 See Julian Raby, 'Pride and prejudice: Mehmed the conqueror and the Italian portrait medal', *Studies in the History of Art*, 21 (1987), pp. 171–94.

10 Julian Raby, 'East and west in Mehmed the conqueror's library', *Bulletin du Bibliophile*, 3 (1987), pp. 297–321: p. 302.

11 On Francesco's attempts to purchase horses from Bayezid, see D. S. Chambers and J. Martineau, eds, *Splendours of the Gonzaga*, exhibition catalogue (London, 1981), p. 147.

12 Cited in G. R. B. Richards, ed., *Florentine Merchants in the Age of the Medici* (Massachusetts, 1932), pp. 163–64.

13 On Cem see J. M. Rogers, ' "The gorgeous east": trade and tribute in the Islamic empires', in *Circa 1492: Art in the Age of Exploration*, ed. Jay Levenson (New Haven, 1991), pp. 69–74.

14 Skelton, *Francesco Berlinghieri*, p. vii.

15 For a discussion and comment on the printing practices of Berlinghieri's text, see Campbell, *Earliest Printed Maps*, pp. 133–35.

16 For just two studies of this aspect of Portuguese navigation and astronomy, see Owen Gingerich, 'Islamic astronomy', *Scientific American*, 254 (1986), pp. 74–83, and Bernard Goldstein, 'The survival of Arabic astronomy in Hebrew', in *Theory and Observation in Ancient and Medieval Astronomy* (London, 1985), ch. 21.

17 See Swerdlow, 'The recovery of the exact sciences'. On Islamic science more generally, see Donald Hill, *Islamic Science and Engineering* (Edinburgh, 1993).

18 Swerdlow, 'The recovery of the exact sciences', p. 145.

19 Dilke, *Greek and Roman Maps*, p. 156.

20 Ahmet Karamustafa, 'Introduction to Islamic Maps', in *The History of Cartography: Cartography in the Traditional Islamic and South Asian Societies*, II, Book 1, eds J. B. Harley and D. Woodward (Chicago, 1992), pp. 3–11: p. 10.

21 Rogers, ' "The gorgeous east" ', p. 70.

22 Necipoğlu, *Architecture, Ceremonial and Power*, pp. 10–15.

23 Cited in Özdemir, *Ottoman Nautical Charts*, p. 52.

24 Michael Kritovoulos, *History of Mehmed the Conqueror* (New Haven, 1954), p. 210.

25 On Donnus Nicolaus, see entry 127 in Levenson (ed.), *Circa 1492*.

26 It is significant that the unfortunate, but astute, Berlinghieri sent manuscript copies of his *Geographia* to Istanbul. On homologies between Italian and Ottoman practices of patronage, and the continuing importance of manuscript production in the early decades of print culture, see Lisa Jardine, *Worldly Goods: A New History of the Renaissance* (London, 1996), pp. 37–132. On Mehmed's extensive patronage of manuscripts and the arts, see Julian Raby, 'A sultan of paradox: Mehmed the Conqueror as a patron of the arts', *Oxford Art Journal*, 5, 1 (1982), pp. 3–8.

27 Cited in Raby, 'Pride and prejudice', p. 187.

28 Franz Babinger, *Mehmed the Conqueror and His Time* (Princeton, 1978), pp. 201–2; Raby, 'East and west', p. 200.

29 Franz Babinger, 'An Italian map of the Balkans, presumably owned by Mehmed the Conqueror', *Imago Mundi*, 8 (1951), pp. 8–15.

30 Campbell, 'Portolan charts', pp. 371–463.

31 Svat Soucek, 'Islamic charting in the Mediterranean', in *The History of Cartography: Cartography in the Traditional Islamic and South Asian Societies*, II, Book 1, eds. J. B. Harley and D. Woodward (Chicago, 1992), pp. 263–92.

32 *Ibid.*, p. 264.

33 Campbell, 'Portolan charts', p. 437. On the hybrid mix of linguistic and navigational concepts employed throughout the eastern Mediterranean, see Henry Kahane and Andreas Tietze, *The Lingua Franca in the Levant* (Illinois, 1958).

34 Brummett, *Ottoman Seapower and Levantine Diplomacy*, pp. 91–97.

35 Piri Reis, *Kitāb-i baḥriye*, I (Istanbul, 1988), p. 43. Little is actually known of the identity of Piri Reis; even the name 'Piri Reis', roughly translated, simply means 'sea captain'.

36 Cited in Özdemir, *Ottoman Nautical Charts*, pp. 60–61.

37 Piri Reis, *Kitāb-i baḥriye*, I, pp. 97–99.

38 *Ibid.*, pp. 107–09.

39 *Ibid.*, pp. 43–45.

40 *Ibid.*, p. 165.

41 Andrew Hess, 'Piri Reis and the Ottoman response to the voyages of discovery', *Terrae Incognitae*, 6 (1974), pp. 19–37.

42 Piri Reis, *Kitāb-i baḥriye*, p. 141.

43 Such an argument would seem to supplement, rather than contradict, the highly influential work of the French historian Fernand Braudel. See his *The Mediterranean and the Mediterranean World in the Age of Philip II* (London, 1992).

44 On the diverse range of contemporary Ottoman topographical maps and plans see Ahmet Karamustafa, 'Military, administrative, and scholarly maps and plans', in *The History of Cartography: Cartography in the Traditional Islamic and South Asian Societies*, II, Book 1, eds J. B. Harley and D. Woodward (Chicago, 1992), pp. 209–227. See also Gülru Necipoğlu, 'Plans and models in 15th- and 16th-century Ottoman architectural practice', *Journal of the Society of Architectural Historians*, 45 (1986), pp. 224–43.

45 On Maṭrākçī, see Dominique Halbout du Tanney, *Istanbul Seen by Maṭrākçī and the Miniatures of the Sixteenth Century* (Istanbul, 1996).

46 See Thomas Goodrich, 'The earliest Ottoman maritime atlas – the Walters *Deniz Atlasi*', *Archivum Ottomanicum*, 11 (1988), pp. 25–50, and Soucek, 'Islamic charting in the Mediterranean', pp. 279–84.

47 Özdemir, *Ottoman Nautical Charts*, pp. 105–8.

48 Karamustafa, 'Introduction to Islamic Maps', p. 225. Nor is this reductive account of manuscript cultures active within the first age of printing limited to other cultures; European manuscript culture remained active elements of the intellectual world of sixteenth-century culture. On this aspect of manuscript culture see Anthony Grafton and Lisa Jardine, ' "Studied for action": how Gabriel Harvey read his Livy', *Past and Present*, 129 (1990), pp. 30–78.

49 Edward Said, 'East isn't east', *Times Literary Supplement* (February 1995), pp. 3–6: p. 6.

50 For a particularly convincing account of the ways in which the figure of the 'despotic' Turk is emplotted across sixteenth-century perceptions of the Ottoman empire from the perspective of later Enlightenment accounts of Ottoman political authority, see Lucette Valensi, 'The making of a political paradigm: the Ottoman state and oriental despotism', in *The Transmission of Culture in Early Modern Europe*, eds Anthony Grafton and Ann Blair (Philadelphia, 1990), pp. 173–203.

4 *Cunning Cosmographers: Mapping the Moluccas*

1 Diffie and Winius, *Foundations of the Portuguese Empire*, pp. 317–22. For a more recent account of the development of the spice trade from the early sixteenth century onwards, see James Boyajian, *Portuguese Trade in Asia under the Habsburgs, 1580–1640* (Baltimore, 1993).

2 Donald Lach, *Asia in the Making of Europe*, I, p. 109.

3 Cited in Diffie and Winius, *Foundations of the Portuguese Empire*, p. 256.

4 Parry, *The Age of Reconnaissance*, p. 201.

5 Diffie and Winius, *Foundations of the Portuguese Empire*, p. 364. On Magellan, see Visconde de Lagoâ, *Fernão de Magalhães (a sua vida e a sua viagem)*, 2 vols (Lisbon, 1938).

6 Cited in Carlos Quirino, ed., *First Voyage Around the World by Antonio Pigafetta and 'De Moluccis Insulis' by Maximilianus Transylvanus* (Manila, 1969), pp. 112–13.

7 Cited in Samuel Eliot Morison, *The European Discovery of America: The Southern Voyages, 1492–1616* (Oxford, 1974), p. 324.

8 Antonio Pigafetta, *Magellan's Voyage. A Narrative Account of the First Circumnavigation* (New Haven, 1969), I, p. 51. Whilst it is unclear as to whether or not Magellan actually consulted Behaim's terrestrial globe, as reference is made here to a 'chart', there seems little doubt that whatever maps and charts (which have not survived) Behaim constructed were based on similar geographical information to that applied in the construction of his globe. Magellan was therefore inheriting a view of the world which must have looked very much like the shape of the terrestrial sphere represented by the Behaim globe.

9 Marcel Destombes, 'The chart of Magellan', *Imago Mundi*, 12 (1955), pp. 65–79.

10 Bartholomew Leonardo de Argensola, *The Discovery and Conquest of the Molucco Islands* (London, 1708), p. 11.

11 Cited in Destombes, 'Chart of Magellan', p. 66.

12 Morison, *European Discovery of America*, p. 343.

13 Destombes, 'Chart of Magellan', p. 68.

14 Lach, *Asia in the Making of Europe*, I, pp. 122–23. On the commercial conflicts which emerged from the Diet, see the discussion in Léon Schick, *Jacob Fugger: un grand homme d'affaires au début du XVIe siècle* (Paris, 1957).

15 Cited in Quirino, *First Voyage Around the World*, p. 34.

16 Cited in Julia Cartwright, *Isabella d'Este, Marchioness of Mantua 1474–1539: A Study of the Renaissance* (London, 1903), II, pp. 225–26.

17 Morison, *European Discovery of America*, pp. 343–44.

18 Pigafetta, *Magellan's Voyage*, p. 67.

19 *Ibid*., p. 68.

20 Cited in Morison, *European Discovery of America*, p. 472.

21 Diffie and Winius, *Foundations of the Portuguese Empire*, p. 283.

22 Davenport, *European Treatises*, p. 187.

23 *Ibid*., p. 187.

24 Cited in Destombes, 'Chart of Magellan', p. 78.

25 Richard Eden, *The Decades of the Newe World* (London, 1555), p. 241.

26 L. A. Vigneras, 'The cartographer Diogo Ribeiro', *Imago Mundi*, 16 (1962), pp. 76–83: p. 76.

27 *Ibid*., p. 77.

28 Eden, *Decades of the Newe Worlde*, p. 242. For details of the negotiations, also see Raimundo de Bulhão Pato, ed., *Cartas de Affonso de Albuquerque seguidas de documentos que as elucidam*, 7 vols (Lisbon, 1884–1935), IV, pp. 73–173.

29 Davenport, *European Treatises*, p. 188.

30 Diffie and Winius, *Foundations of the Portuguese Empire*, p. 283.

31 Cited in Davenport, *European Treatises*, p. 161.

32 *Ibid*., p. 188. Emphasis added.

33 Edward Heawood, 'The world map before and after Magellan's voyage', *Geographical Journal*, 57 (1921), pp. 431–42: pp. 431–32. Whilst debate continued to rage over competing claims to the circumference of the earth, based on both classical and more contemporary calculations, as the ensuing debates at Saragossa emphasize, Ptolemy was consistently used as the ultimate arbiter.

34 *Ibid*., p. 432.

35 Henry Wagner, *Spanish Voyages to the Northwest Coast of America in the Sixteenth Century* (Berkeley, 1929), p. 95.

36 Heawood, 'The world map', p. 435.

37 *Ibid*., p. 435.

38 E. G. R. Taylor, ed., *A Brief Summe of Geographie by Robert Barlow* (London, 1932), pp. xi–xx.

39 *Ibid.*, pp. xvi–xvii.

40 Argensola, *Discovery and Conquest*, p. 31.

41 Cited in E. H. Blair and J. A. Robertson, eds, *The Philippine Islands, 1493–1803*, (Cleveland, 1903), I, p. 198.

42 *Ibid.*, pp. 209–10.

43 Destombes, 'Chart of Magellan', p. 69.

44 Shirley, *The Mapping of the World*, p. xxv.

45 Crone, *Maps and their Makers*, p. 95.

46 Heawood, 'The world map', p. 441.

47 Lawrence C. Wroth, 'The early cartography of the Pacific', *Papers of the Bibliographical Society of America*, 38 (1944), pp. 187–267: p. 251.

48 Arjun Appadurai, ed., *The Social Life of Things: Commodities in Cultural Perspective* (Cambridge, 1986), pp. 3–63.

49 Cited in Rawdon Brown, ed., *Calendar of State Papers: Venetian* (London, 1871), IV, p. 209.

50 See Karrow, *Mapmakers of the Sixteenth Century*, pp. 206–7, and Robert Haardt, 'The globe of Gemma Frisius', *Imago Mundi*, 11 (1952), pp. 109–10.

51 Cited in van der Krogt, *Globi Neerlandici*, p. 76.

52 Argensola, *Discovery and Conquest*, p. 30.

5 *Plotting and Projecting: The Geography of Mercator and Ortelius*

1 Robert Thorne, 'A Declaration of the Indies', in Richard Hakluyt, *Divers Voyages Touching America* (London, 1582), sig. c3.

2 Chrisman, *Lay Culture, Learned Culture*, p. 138.

3 Hildegard Johnson, *Carta Marina: World Geography in Strassburg, 1525* (Minnesota, 1963), p. 124.

4 Cited in Haardt, 'Gemma Frigius', pp. 109–10.

5 Cited in van der Krogt, *Globi Neerlandici*, p. 54.

6 M. Van Durme, ed., *Correspondance Mercatorienne* (Antwerp, 1959), p. 18. On Mercator's output, see F. Van Ortroy, *Bibliographie de l'Oeuvre Mercatorienne* (Amsterdam, 1978).

7 Van Durme, *Correspondance Mercatorienne*, p. 14.

8 Walter Ghim, 'Vita Mercatoris', in *Mercator*, ed. A. S. Osley (London, 1969), pp. 185–94: p. 186.

9 Cited in van der Krogt, *Globi Neerlandici*, p. 60.

10 *Ibid.*, p. 70.

11 Helen Wallis, 'Intercourse with the peaceful muses', in *Across the Narrow Seas. Studies in the History and Bibliography of Britain and the Low Countries*, ed. Susan Roach (London, 1991), pp. 31–54: p. 34.

12 Van Durme, *Correspondance Mercatorienne*, p. 35.

13 Ghim, 'Vita Mercatoris', pp. 186–87.

14 Gerard Mercator, *Gerard Mercator's Map of the World (1569)* (Rotterdam, 1961), p. 18.

15 *Ibid.*, p. 45.

16 *Ibid.*, p. 46.

17 *Ibid.*, p. 46.

18 *Ibid.*, p. 46.

19 On this historically later development of perceptions of the 'east', see Abou-el-Haj, 'The formal closure', and Larry Wolff, *Inventing Eastern Europe* (Stanford, 1994).

20 Ir. C. Koeman, *Atlantes Neerlandici* (Amsterdam, 1969), II, p. 25.

21 Ir. C. Koeman, *The History of Abraham Ortelius and his 'Theatrum Orbis Terrarum'* (Lausanne, 1964), p. 13.

22 Koeman, *Atlantes Neerlandici*, III, p. 26.
23 Hessels, *Ortelii Epistulae*, no. 330.
24 Koeman, *History*, pp. 38–39.
25 Hessels, *Ortelii Epistulae*, no. 40.
26 *Ibid.*, no. 32. For details of Ortelius' maps, see Marcel van den Broecke, *Ortelius Atlas Maps* (Netherlands, 1996).
27 Hessels, *Ortelii Epistulae*, no. 86.
28 Abraham Ortelius, 'Address to the Reader', in *Theatrum Orbis Terrarum* (London, 1606), unpaginated.
29 *Ibid.*
30 *Ibid.*
31 *Ibid.*
32 Eisenstein, *The Printing Press*, p. 519. See also Karrow, *Mapmakers of the Sixteenth Century.*
33 Cited in Koeman, *History*, p. 36.
34 *Ibid.*, p. 18.
35 *Ibid.*, p. 18
36 More recently a range of critics and historians of cartography have questioned the validity of the truth claims made on behalf of modern, 'objective' cartography. See for instance the excellent studies of J. B. Harley, 'Maps, knowledge and power', in *The Iconography of Landscape*, eds Denis Cosgrove and Stephen Daniels (Cambridge, 1988), pp. 277–312, and 'Deconstructing the map', in *Writing Worlds: Discourse, Text and Metaphor in the Representation of Landscape*, eds Trevor Barnes and James Duncan (London, 1992), pp. 231–47. See also Denis Wood, *The Power of Maps* (New York, 1992).

6 *Conclusion*

1 Cited in George Bruner Parks, *Richard Hakluyt and the English Voyages* (New York, 1928), p. 155.
2 *Ibid.*, p. 155.
3 Gunter Schilder, 'Organization and evolution of the Dutch East India Company's hydrographic office in the seventeenth century', *Imago Mundi*, 28 (1976), pp. 61–78.
4 Ir. C. Koeman, *Joan Blaeu and his Grand Atlas* (Amsterdam, 1970).

Bibliography

Rifaat A. Abou-el-Haj, 'The formal closure of the Ottoman frontier in Europe, 1699–1703', *Journal of the American Oriental Society*, 89 (1969), pp. 467–75.

Luís de Albuquerque, *Curso de História da Nautica*, Coimbra, 1972.

—— *Historia de la navegación portuguesa*, Madrid, 1991.

Robert Almagia and Tullia Gasparrini, *Il Mappamondo di Fra Mauro*, Florence, 1956.

Fortunato de Almeida, *O Infante de Sagres*, Oporto, 1894.

Arjun Appadurai, ed., *The Social Life of Things: Commodities in Cultural Perspective*, Cambridge, 1986.

Bartholomew Leonardo de Argensola, *The Discovery and Conquest of the Molucco Islands*, London, 1708.

Eric Axelson, ed., *Dias and His Successors*, Cape Town, 1988.

Gomes Eannes de Azurara, *The Chronicle of the Discovery and Conquest of Guinea*, London, 1896.

Franz Babinger, 'An Italian map of the Balkans, presumably owned by Mehmed the Conqueror', *Imago Mundi*, 8 (1951), pp. 8–15.

—— *Mehmed the Conqueror and His Time*, Princeton, 1978.

Leo Bagrow, *The History of Cartography*, London, 1966.

Peter Barber, 'Visual encyclopaedias: the Hereford and other Mappae Mundi', *The Map Collector*, 48 (1989), pp. 2–8.

Charles Raymond Beazley, *Prince Henry the Navigator*, New York, 1923.

Joaquim Bensaúde, *Histoire de la science nautique portugaise a l'époque des grandes découvertes*, 7 vols, Lisbon, 1914–24.

C. F. Black, ed. *et al.*, *Atlas of the Renaissance*, Amsterdam, 1993.

E. H. Blair and J. A. Robertson, eds, *The Philippine Islands, 1493–1803*, I, Cleveland, 1903.

John Blake, *European Beginnings in West Africa, 1457–1578*, London, 1937.

—— ed., *Europeans in West Africa, 1450–1560*, I, London, 1942.

C. R. Boxer, *The Portuguese Seaborne Empire, 1415–1825*, New York, 1969.

James Boyajian, *Portuguese Trade in Asia under the Habsburgs, 1580–1640*, Baltimore, 1993.

José de Braganza, ed., *Crónica do descobrimento e conquista da Guiné*, 2 vols, Oporto, 1937.

Fernand Braudel, *The Mediterranean and the Mediterranean World in the Age of Philip II*, London, 1992.

British Museum, *A Map of the World, Designed by Giovanni Matteo Contarini*, London, 1926.

José Brocahdo, *O piloto Árabe de Vasco da Gama*, Lisbon, 1959.

Marcel van den Broecke, *Ortelius Atlas Maps*, Netherlands, 1996.

Lloyd Brown, *The Story of Maps*, New York, 1949.

Rawdon Brown, ed., *Calendar of State Papers: Venetian*, IV, London, 1871.

Palmira Brummett, *Ottoman Seapower and Levantine Diplomacy in the Age of Discovery*, New York, 1994.

David Buisseret, ed., *Monarchs, Ministers and Maps: The Emergence of Cartography as a Tool of Government in Early Modern Europe*, Chicago, 1992.

Raimundo de Bulhão Pato, ed., *Cartas de Affonso de Albuquerque seguidas de documentos que as elucidam*, 7 vols, Lisbon, 1884–1935.

Jacob Burckhardt, *The Civilisation of the Renaissance in Italy*, London, 1892.

Tony Campbell, *The Earliest Printed Maps, 1472–1500*, London, 1987.

—— 'Portolan charts from the late thirteenth century to 1500', in *The History of Cartography: Cartography in Prehistoric, Ancient, and Mediaeval Europe and the Mediterranean*, I, eds J. B. Harley and D. Woodward, Chicago, 1987, pp. 371–463.

Julia Cartwright, *Isabella d'Este, Marchioness of Mantua 1474–1539: A Study of the Renaissance*, 2 vols, London, 1903.

D. S. Chambers and J. Martineau, eds, *Splendours of the Gonzaga*, exhibition catalogue, London, 1981.

K. N. Chaudhuri, *Trade and Civilization in the Indian Ocean: An Economic History from the Rise of Islam to 1750*, Cambridge, 1985.

Miriam Usher Chrisman, *Lay Culture, Learned Culture: Books and Social Change in Strasbourg, 1480–1599*, New Haven, 1982.

Gaspar de Corrêa, ed., *Lendas da Índia*, 8 vols, Lisbon, 1856–66.

Armando Cortesão, *History of Portuguese Cartography*, II, Coimbra, 1969.

—— and Avelino Teixeira da Mota, *Portugaliae monumenta Cartographica*, 6 vols, Coimbra, 1958–63.

G. R. Crone, *Maps and Their Makers*, London, 1953.

—— ed., *The Voyages of Cadamosto and other Documents on Western Africa in the Second Half of the Fifteenth Century*, London, 1937.

William Cunningham, *The Cosmographicall Glasse*, London, 1559.

Frances Gardiner Davenport, ed., *European Treatises on the History of the United States and its Dependencies*, I, Washington, 1967.

John Dee, *Preface to Euclid's 'Elements of Geometrie'*, London, 1570.

Marcel Destombes, 'The chart of Magellan', *Imago Mundi*, 12 (1955), pp. 65–79.

B. Diffie and G. Winius, *Foundations of the Portuguese Empire, 1415–1580*, Minneapolis, 1977.

O. A. W. Dilke, *Greek and Roman Maps*, London, 1985.

—— 'Cartography in the Byzantine empire', in *The History of Cartography: Cartography in Prehistoric, Ancient and Mediaeval Europe and the Mediterranean*, I, eds J. B. Harley and D. Woodward, Chicago, 1987, pp. 258–75.

M. Van Durme, ed., *Correspondance Mercatorienne*, Antwerp, 1959.

Richard Eden, *The Decades of the Newe Worlde*, London, 1555.

Elizabeth Eisenstein, *The Printing Press as an Agent of Change: Communications and Cultural Transformations in Early Modern Europe*, 2 vols, Cambridge, 1979.

Lucien Febvre and Henri-Jean Martin, *The Coming of the Book: The Impact of Printing 1450–1800*, London, 1976.

Felipe Fernández-Armesto, *Before Columbus: Exploration and Colonisation from the Mediterranean to the Atlantic, 1229–1492*, London, 1987.

Focus Behaim-Globus, Nuremberg, 1991.

Grete de Francesco, 'The Venetian silk industry', *Ciba Review* 29 (1940), pp. 1027–35.

John B. Friedman, 'Cultural conflicts in mediaeval world maps', in *Implicit Understandings. Observing, Reporting, and Reflecting on the Encounters Between Europeans and Other Peoples in the Early Modern Era*, ed. Stuart B. Schwartz, Cambridge, 1994, pp. 64–95.

Walter Ghim, 'Vita Mercatoris', in *Mercator*, ed. A. S. Osley, London, 1969, pp. 185–94.

Vitorino Magalhães Godinho, *A economia dos descobrimentos henriquinos*, Lisbon, 1962.

—— *Os descobrimentos e a economia mundial*, 2 vols, Lisbon, 1965–71.

Owen Gingerich, 'Islamic astronomy', *Scientific American*, 254 (1986), pp. 74–83.

Bernard Goldstein, 'The survival of Arabic astronomy in Hebrew', in *Theory and Observation in Ancient and Medieval Astronomy*, London, 1985.

Thomas Goodrich, 'The earliest Ottoman maritime atlas – the Walters *Deniz Atlasi*', *Archivum Ottomanicum*, 11 (1988), pp. 25–50.

Margreta de Grazia, 'The ideology of superfluous things: *King Lear* as period piece', in *Subject and Object in Renaissance Culture*, ed. Margreta de Grazia *et al.*, Cambridge, 1996, pp. 17–42.

Anthony Grafton and Lisa Jardine, ' "Studied for action": how Gabriel Harvey read his Livy', *Past and Present*, 129 (1990), pp. 30–78.

Stephen Greenblatt, *Renaissance Self-Fashioning: From More to Shakespeare*, Chicago, 1980.

—— *Marvelous Possessions: The Wonder of the New World*, Oxford, 1991.

W. B. Greenlee, ed., *The Voyage of Pedro Alvares Cabral to Brazil and India*, London, 1937.

Annemarie Jordan Gschwend, 'In the tradition of princely collections: curiosities and exotica in the *Kunstkammer* of Catherine of Austria', *Bulletin of the Society for Renaissance Studies*, 13, 1 (1995), pp. 1–9.

A. Guimaràes, *Uma corrente do colonialismo Português: a Sociedade de Geografia de Lisboa, 1875–1895*, Lisbon, 1984.

Robert Haardt, 'The globe of Gemma Frisius', *Imago Mundi*, 11 (1952), pp. 109–10.

John Hale, *The Civilization of Europe in the Renaissance*, London, 1994.

Kim Hall, *Things of Darkness: Economies of Race and Gender in Early Modern England*, Cornell, 1995.

Clarence Henry Haring, *Trade and Navigation Between Spain and the Indies*, Massachusetts, 1918.

J. B. Harley, 'Maps, knowledge and power', in *The Iconography of Landscape*, eds Denis Cosgrove and Stephen Daniels, Cambridge, 1988, pp. 277–312.

—— 'Deconstructing the map', in *Writing Worlds: Discourse, Text and Metaphor in the Representation of Landscape*, eds Trevor Barnes and James Duncan, London, 1992, pp. 231–47.

—— *Maps and the Columbian Encounter*, Milwaukee, 1990.

Edward Heawood, 'The world map before and after Magellan's voyage', *Geographical Journal*, 57 (1921), pp. 431–42.

Mary Helms, *Ulysses' Sail: An Ethnographic Odyssey of Power, Knowledge, and Geographical Distance*, Princeton, 1988.

Andrew Hess, 'Piri Reis and the Ottoman response to the voyages of discovery', *Terrae Incognitae*, 6 (1974), pp. 19–37.

J. H. Hessels, ed., *Ortelii Epistulae*, Oxford, 1887.

Wilhelm von Heyd, *Histoire de commerce du Levant au Moyen Age*, 2 vols, Leipzig, 1885–86.

Donald Hill, *Islamic Science and Engineering*, Edinburgh, 1993.

Jonathan Israel, *Dutch Primacy in World Trade, 1585–1740*, Oxford, 1989.

William Ivins, *Prints and Visual Communications*, New York, 1969.

Lisa Jardine, *Worldly Goods: A New History of the Renaissance*, London, 1996.

Hildegard Johnson, *Carta Marina: World Geography in Strassburg, 1525*, Minnesota, 1963.

Henry Kahane and Andreas Tietze, *The Lingua Franca in the Levant*, Illinois, 1958.

Paul Kahle, 'A lost map of Columbus', *Geographical Journal*, 23 (1933), pp. 621–38.

Ahmet T. Karamustafa, 'Introduction to Islamic Maps', in *The History of Cartography: Cartography in the Traditional Islamic and South Asian Societies*, II, Book 1, eds J. B. Harley and D. Woodward, Chicago, 1992, pp. 3–11.

—— 'Military, administrative, and scholarly maps and plans', in *The History of Cartography: Cartography in the Traditional Islamic and South Asian Societies*, II, Book 1, eds J. B. Harley and D. Woodward, Chicago, 1992, pp. 209–227.

Robert Karrow, *Mapmakers of the Sixteenth Century and their Maps: Bio-Bibliographies of the Cartographers of Abraham Ortelius*, Chicago, 1993.

Johannes Keuning, 'The history of geographical map projections until 1600', *Imago Mundi*, 12 (1955), pp. 1–24.

Ir. C. Koeman, *Atlantes Neerlandici*, 3 vols, Amsterdam, 1969.

—— *The History of Abraham Ortelius and His 'Theatrum Orbis Terrarum'*, Lausanne, 1964.

——*Joan Blaeu and his Grand Atlas*, Amsterdam, 1970.

Peter van der Krogt, *Globi Neerlandici. The Production of Globes in the Low Countries*, Utrecht, 1993.

Michael Kritovoulos, *History of Mehmed the Conqueror*, Princeton, 1954.

Donald Lach, *Asia in the Making of Europe*, I, Book 1, Chicago, 1965.

——*Asia in the Making of Europe*, I, Book 2, Chicago, 1970.

Visconde de Lagoâ, *Fernão de Magalhães (a sua vida e a sua viagem)*, 2 vols, Lisbon, 1938.

Jonathan Lanman, 'The Religious Symbolism of the T in T-O Maps', *Cartographica*, 18 (1981), pp. 18–22.

Duarte Leite, *Descobrimentos portuguêses*, 2 vols, Lisbon, 1958–60.

Frank Lestringant, *Mapping the Renaissance World: The Geographical Imagination in the Age of Discovery*, Cambridge, 1994.

Cândido Lusitano, *Vida do Infante D. Henrique*, Lisbon, 1758.

Helder Macedo, 'Recognizing the unknown: perceptions in the age of European expansion', *Portuguese Studies*, 8 (1992), pp. 130–36.

João Franco Machado and Damião Peres, eds, *Viagens de Lúis de Cadamosto e Pedro Sintra*, Lisbon, 1948.

Francis Maddison, 'A consequence of discovery: astronomical navigation in fifteenth-century Portugal', in *Studies in the Portuguese Discoveries I*, eds T. F. Earle and S. Parkinson, Warminster, 1992, pp. 71–110.

L. Maduerira, 'The discreet seductiveness of the crumbling empire', *Luso-Brazilian Review*, 32, 1 (1995), pp. 17–29.

António de Magalhães Basto, ed., *Diário da viagem de Vasco da Gama*, Oporto, 1945.

Alfredo Pinheiro Marques, *Origem e desenvolvimento da cartografia portuguesa na época dos descobrimentos*, Lisbon, 1987.

Gerard Mercator, *Gerard Mercator's Map of the World (1569)*, Rotterdam, 1961.

Samuel Eliot Morison, *The European Discovery of America: The Southern Voyages, 1492–1616*, Oxford, 1974.

Avelina Teixeira da Mota, 'Evolução dos roteiros portugueses durante o século XVI', *Revista da Universidade de Coimbra*, 24 (1969), pp. 1–32.

—— 'Some notes on the organization of hydrographical services in Portugal before the nineteenth century', *Imago Mundi*, 28 (1976), pp. 51–60.

Gülru Necipoğlu, 'Plans and models in 15th- and 16th-century Ottoman architectural practice', *Journal of the Society of Architectural Historians*, 45 (1986), pp. 224–43.

—— 'Süleyman the Magnificent and the representation of power in the context of Ottoman-Habsburg-Papal rivalry', *Art Bulletin*, 71 (1989), pp. 401–27.

—— *Architecture, Ceremonial and Power: The Topkapi Palace in the Fifteenth and Sixteenth Centuries*, Massachusetts, 1991.

Adolf Nordenskiöld, *Facsimile-Atlas to the Early History of Cartography with Reproductions of the Most Important Maps Printed in the XV and XVI Centuries*, New York, 1973.

Abraham Ortelius, *Theatrum Orbis Terrarum*, London, 1606.

F. Van Ortroy, *Bibliographie de l'Oeuvre Mercatorienne*, Amsterdam, 1978.

Kemal Özdemir, *Ottoman Nautical Charts and the Atlas of 'Ali Mācār Reis*, Istanbul, 1992.

Sergio Pacifici, ed., *Copy of a Letter of the King of Portugal Sent to the King of Castile Concerning the Voyage and Success in India*, Minneapolis, 1955.

George Bruner Parks, *Richard Hakluyt and the English Voyages*, New York, 1928.

J. H. Parry, *The Age of Reconnaissance: Discovery, Exploration and Settlement, 1450–1650*, London, 1973.

—— *Europe and a Wider World, 1415–1715*, London, 1949.

Boies Penrose, *Travel and Discovery in the Renaissance, 1420–1620*, Massachusetts, 1952.

Duarte Pacheco Pereira, *Esmeraldo de Situ Orbis*, London, 1937.

Damião Peres, *História dos descobrimentos portuguêses*, Oporto, 1960.

Antonio Pigafetta, *Magellan's Voyage: A Narrative Account of the First Circumnavigation*, 2 vols, New Haven, 1969.

Claudius Ptolemy, *The Geography*, New York, 1991.

D. B. Quinn, ed., *The Hakluyt Handbook*, I, London, 1974.

Carlos Quirino, ed., *First Voyage Around the World by Antonio Pigafetta and 'De Moluccis Insulis' by Maximilianus Transylvanus*, Manila, 1969.

Julian Raby, 'A sultan of paradox: Mehmed the Conqueror as a patron of the arts', *Oxford Art Journal*, 5, 1 (1982), pp. 3–8.

—— 'East and west in Mehmed the Conqueror's library', *Bulletin du Bibliophile*, 3 (1987), pp. 297–321.

—— 'Pride and prejudice: Mehmed the Conqueror and the Italian portrait medal', *Studies in the History of Art*, 21 (1987), pp. 171–94.

W. G. L. Randles, 'The alleged nautical school founded in the fifteenth century at Sagres by Prince Henry of Portugal, called "The Navigator" ', *Imago Mundi*, 45 (1993), pp. 20–28.

E. G. Ravenstein, ed., *A Journal of the First Voyage of Vasco da Gama, 1497–1499*, London, 1898.

—— *Martin Behaim: His Life and His Globe*, London, 1908.

Piri Reis, *Kitāb-i baḥriye*, 2 vols, Istanbul, 1988.

G. R. B. Richards, ed., *Florentine Merchants in the Age of the Medici*, Massachusetts, 1932.

Francis M. Rogers, ed., *The Obedience of a King of Portugal*, Minneapolis, 1958.

—— *The Travels of the Infante Dom Pedro of Portugal*, Massachusetts, 1961.

J. M Rogers, ' "The gorgeous east": trade and tribute in the Islamic empires', in *Circa 1492: Art in the Age of Exploration*, ed. Jay Levenson, New Haven, 1991, pp. 69–74.

Edward Said, *Orientalism*, London, 1978.

—— 'East isn't east', *Times Literary Supplement* (February 1995), pp. 3–6.

Elaine Sanceau, *Henry the Navigator*, New York, 1947.

Léon Schick, *Jacob Fugger: un grand homme d'affaires au début du XVIe siècle*, Paris, 1957.

Gunter Schilder, 'Organization and evolution of the Dutch East India Company's hydrographic office in the seventeenth century', *Imago Mundi*, 28 (1976), pp. 61–78.

Robert Schwoebel, *The Shadow of the Crescent: The Renaissance Image of the Turk, 1453–1517*, Nieuwkoop, 1967.

Patricia Seed, *Ceremonies of Possession in Europe's Conquest of the New World 1492–1640*, Cambridge, 1995.

Rodney W. Shirley, *The Mapping of the World: Early Printed Maps, 1472–1700*, London, 1983.

R. A. Skelton, *Explorer's Maps*, London, 1958.

—— 'The early map printer and his problems', *Penrose Annual*, 57 (1964), pp. 171–87.

—— ed., *Francesco Berlinghieri: Geographia*, Amsterdam, 1966.

—— 'A contract for world maps at Barcelona, 1399–1400', *Imago Mundi*, 22 (1968), pp. 107–13.

Svat Soucek, 'Islamic charting in the Mediterranean', in *The History of Cartography: Cartography in the Traditional Islamic and South Asian Societies*, II, Book 1, eds J. B. Harley and D. Woodward, Chicago, 1992, pp. 263–92.

Edward Luther Stevenson, *Celestial and Terrestrial Globes*, I, New Haven, 1921.

N. M. Swerdlow, 'The recovery of the exact sciences of antiquity: mathematics, astronomy, geography', in *Rome Reborn: The Vatican Library and Renaissance Culture*, ed. Anthony Grafton, Washington, 1993, pp. 125–67.

Dominique Halbout du Tanney, *Istanbul Seen by Maṭrākçī and the Miniatures of the Sixteenth Century*, Istanbul, 1996.

E. G. R. Taylor, ed., *A Brief Summe of Geographie by Robert Barlow*, London, 1932.

—— ed., *The Original Writings and Correspondence of the Two Richard Hakluyts*, London, 1935.

—— *The Haven-finding Art: A History of Navigation from Odysseus to Captain Cook*, London, 1956.

Robert Thorne, 'A Declaration of the Indies', in Richard Hakluyt, *Divers Voyages Touching America*, London, 1582.

Gerald R. Tibbetts, 'The role of charts in Islamic navigation in the Indian Ocean', in *The History of Cartography: Cartography in the Traditional Islamic and South Asian Societies*, II, Book 1, eds J. B. Harley and D. Woodward, Chicago, 1992, pp. 256–62.

Ronald Tooley, *Maps and Map-Makers*, London, 1949.

Norman Thrower, *Maps and Civilization: Cartography in Culture and Society*, Chicago, 1996.

David Turnbull, 'Cartography and science in early modern Europe: mapping the construction of knowledge spaces', *Imago Mundi*, 48 (1996), pp. 5–24.

Lucette Valensi, 'The making of a political paradigm: the Ottoman state and oriental despotism', in *The Transmission of Culture in Early Modern Europe*, eds Anthony Grafton and Ann Blair, Philadelphia, 1990, pp. 173–203.

L. A. Vigneras, 'The cartographer Diogo Ribeiro', *Imago Mundi*, 16 (1962), pp. 76–83.

Henry R. Wagner, *Spanish Voyages to the Northwest Coast of America in the Sixteenth Century*, Berkeley, 1929.

Martin Waldseemüller, *The Cosmographiae Introductio*, New York, 1907.

Helen Wallis, 'Intercourse with the peaceful muses', in *Across the Narrow Seas. Studies in the History and Bibliography of Britain and the Low Countries*, ed. Susan Roach, London, 1991, pp. 31–54.

David Waters, 'Science and the techniques of navigation in the Renaissance', in *Art, Science and History in the Renaissance*, ed. Charles Singleton, Baltimore, 1967, pp. 189–237.

Donald Weinstein, ed., *Ambassador From Venice: Pietro Pasqualigo in Lisbon, 1501*, Minneapolis, 1960.

F. C. Wieder, *Monumenta Cartographica*, I, The Hague, 1925–33.

—— *De Oude Weg Naar Indië Ons De Kaap*, Amsterdam, 1915.

Larry Wolff, *Inventing Eastern Europe*, Stanford, 1994.

Denis Wood, *The Power of Maps*, New York, 1992.

David Woodward, 'Maps and the rationalization of geographic space', in *Circa 1492: Art in the Age of Exploration*, ed. Jay Levenson, New Haven, 1991, pp. 83–88.

—— 'Medieval *Mappaemundi*', in *The History of Cartography: Cartography in the Prehistoric, Ancient and Mediaeval Europe and the Mediterranean*, I, eds J. B. Harley and D. Woodward, Chicago, 1987, pp. 286–370.

Lawrence C. Wroth, 'The early cartography of the Pacific', *Papers of the Bibliographical Society of America*, 38 (1944), pp. 187–267.

Index

204